Global Standard-Setting in Internet Governance

Global Standard-Setting in Internet Governance

By
ALISON HARCOURT
University of Exeter

GEORGE CHRISTOU
University of Warwick

SEAMUS SIMPSON
University of Salford

OXFORD
UNIVERSITY PRESS

OXFORD
UNIVERSITY PRESS

Great Clarendon Street, Oxford, OX2 6DP,
United Kingdom

Oxford University Press is a department of the University of Oxford.
It furthers the University's objective of excellence in research, scholarship,
and education by publishing worldwide. Oxford is a registered trade mark of
Oxford University Press in the UK and in certain other countries

First Edition published in 2020

Impression: 1

Published in the United States of America by Oxford University Press
198 Madison Avenue, New York, NY 10016, United States of America

British Library Cataloguing in Publication Data
Data available

Library of Congress Control Number: 2019947761

ISBN 978-0-19-884152-4

Printed and bound by
CPI Group (UK) Ltd, Croydon, CR0 4YY

Acknowledgements

We would like to thank the many organizations, experts, and institutions that facilitated the production of this book. First we would like to thank Natalie Patey, Emma Taylor, and Alex Flach at Oxford University Press for their commitment and support. We would also like to thank the external reviewers for insightful comments and suggestions for the monograph.

Second, we thank the Economic and Social Research Council (ESRC) which funded our project, International Professional Fora: A Study of Civil Society Organisation Participation in Internet Governance, which ran between September 2015 and August 2018. We are grateful to our project team members Imir Rashid and Francesca Farmer at the University of Exeter for execution of research and activities related to this book. We are grateful to our project external steering committee members: Diane Stone at the University of Warwick; Nikolaos Zahariadis at Rhodes College; Megan Richards at the European Commission; Jean Francois Furnemont from the European Platform for Regulatory Authorities; Jim Killock, Executive Director of the Open Rights Group; Rigo Wenning, Legal Counsel for the W3C; and Pieter de Wit, formerly Director of Community Media Forum Europe and member of Pascal Lamy's European Commission advisory group on the future use of UHF spectrum for wireless broadband and TV.

The ESRC project funded a number of workshops in April 2017 and May 2018 at the University of Warwick office in Brussels and in November 2017 at Portcullis House in London. We would like to thank the many experts that gave presentations at these workshops for providing us with their knowledge, time, and insight into standard-setting. We would also like to thank our interviewees, the many policy experts, government, and public policy offices, and civil society representatives for their time and expertise.

We thank the many administrative staff members at the Universities of Exeter, Salford, and Warwick for their support and also James Loud at 146Webdesign. Alison Harcourt would like to thank Martin Lodge and the Centre for Analysis of Risk and Regulation at the London School of Economics for hosting her in 2018 and 2019. Finally, we would like to thank our respective partners and families for their support.

Acknowledgements

Contents

List of Tables and Figures

Table of Cases

EUROPEAN UNION

UNITED STATES

Table of Legislations

INTERNATIONAL

List of Abbreviations

3GPP	Third Generation Partnership Project
AAP	Association of American Publishers
ACAP	Automated Content Access Protocol
ACC	Advanced Audio Coding
ACLU	American Civil Liberties Union
ACME	Automated Certificate Management Environment
ACTA	Anti-Counterfeiting Trade Agreement
AD	Area Directors (IETF)
ADC	Asociación por los Derechos Civiles
AEF	Agricultural Industry Electronics Foundation
AFNOR	Association Française de Normalisation
ALTO	Application-Layer Traffic Optimization
ANSI	American National Standards Institute
APA	Accessible Platform Architectures
APC	Association for Progressive Communications
API	Application programming interface
APP	Atom Publishing Protocol
ARPANET	Advanced Research Projects Agency Network
ASD	Australian Signals Directorate
AWS	Amazon Web Services
BBC	British Broadcasting Corporation
BCAUSE	Broadband Network Gateway Control-Plane And User-Plane Separation
BEIRG	British Entertainment Industry Group
BGP	Border Gateway Protocol
BIMI	Brand Indicators for Message Identification
BIS	Bank for International Settlements
BoD	Board of Directors
BOF	Birds of a Feather
BPI	British Phonographic Industry
BRAN	Broadband Random Access Networks
BSA	British Software Alliance
CA	certificate authority
CACAO	Collaborative Automated Course of Action Operations
CALEA	Communication Assistance for Law Enforcement Act
CalOPPA	California Online Privacy Protection Act
CAS	conditional access systems
CCPA	California Consumer Privacy Act

CCXML	Call Control eXtensible Markup Language
CDM	Content Decryption Module
CDN	Content Delivery Network
CDP	centralized differential privacy
CDT	Centre for Democracy and Technology
CEN	European Committee for Standardization (Comité Européen de Normalisation)
CENELC	European Committee for Electrotechnical Standardization (Comité Européen de Normalisation Électrotechnique)
CEPT	European Conference of Postal and Telecommunications Administrations
CERN	European Organization for Nuclear Research
CESG	Communications-Electronics Security Group
CIA	Central Intelligence Agency
CIS	Centre for Internet and Society
CISAC	International Confederation of Societies of Authors and Composers (Confédération Internationale des Sociétés d'Auteurs et Compositeurs)
CMRS	Commercial Mobile Radio Services
CNiL	French Commission nationale de l'informatique et des libertés
CNRI	Corporation for National Research Initiatives
CoE	Council of Europe
CSAIL	Computer Science and Artificial Intelligence Laboratory
CSD	Criteria for Standards Development
CSEC	Canadian Communications Security Establishment
CSMA/CA	carrier sense multiple access with collision avoidance
CSS	Cascading Style Sheet
CSO	Civil society organization
CT	Certificate Transparency
CTAP	Client to Authenticator Protocol
DAA	Digital Advertising Alliance
DANE	DNS-based Authentication of Named Entities
DARPA	Defense Advanced Research Projects Agency
DES	Data Encryption Standard
DIN	Deutsches Institut für Normung
DKIM	DomainKeys Identified Mail
DMARC	Domain-based Message Authentication, Reporting & Conformance
DMCA	Digital Millennium Copyright Act
DOI	Digital Object Identifier
DOM	Document Object Model
DNS	Domain Name System
DNT	Do Not Track
DP	differential privacy
DRM	Digital rights management
DSA	Dynamic Spectrum Access

DSC	Dynamic Sensitivity Control
DSL	Digital Subscriber Line
DTT	Digital Terrestrial Television
DUAL_EC_DRBG	Dual Elliptic Curve Deterministic Random Bit Generator
DVB	Digital Video Broadcasting
ECMA	European Computer Manufacturer's Association
ECTRA	European Committee for Telecommunications Regulatory Affairs
EDRi	European Digital Rights initiative
EFF	Electronic Frontier Foundation
EME	Encrypted Media Extensions
EMMA	Extensible MultiModal Annotation markup language
ENISA	European Union Agency for Network and Information Security
EPAG	EPAG Domainservices GmbH
EPC	European Publishers Council
EPP	Extensible Provisioning Protocol
ePR	ePrivacy Regulation
ESA	Encode, Shuffle, Analyze
ESC	Economic and Social Committee
ESO	European Standards Organization
ETNO	European Telecommunications Network Operators' Association
ETSI	European Telecommunications Standards Institute
EU	European Union
FCC	Federal Communications Commission
FIDO	Fast IDentity Online
FIPS	Federal Information Processing Standard
FISA	Foreign Intelligence Surveillance Act
FPF	Future of Privacy Forum
FRAND	Fair, reasonable, and non-discriminatory
FSF	Free Software Foundation
FTA	Free trade agreement
FTC	Federal Trade Commission
FTP	File Transfer Protocol
GCHQ	Government Communications Headquarters
GCSB	Government Communications Security Bureau (New Zealand)
GDPR	General Data Protection Regulation
GEOPRIV	Geographic Location/Privacy
GIF	Graphics Interchange Format
GKP	Global Knowledge Partnership
GPG	GNU Privacy Guard
GTP	General Packet Radio Services Tunneling Protocol
GSA	General Services Administration (US)
GSM	Global System for Mobile Communications
gTLD	generic top-level domain name
HCSEC	Huawei Cyber Security Evaluation Centre
HDTV	high-definition television

HRPC	Human Rights Protocol Considerations Research Group
HTML	Hypertext Markup Language
HTTP	Hypertext transfer protocol
IAB	International Advertising Bureau
IAB	Internet Architecture Board
IANA	Internet Assigned Numbers Authority
IAOC	Internet Administrative Oversight Committee
ICANN	Internet Corporation for Assigned Names and Numbers
ICCPR	International Covenant on Civil and Political Rights
ICMP	Internet Control Message Protocol
ICT	Information and Communication Technologies
ICTSB	Information and Communication Technology Standards Board
IE	Internet Explorer
IEC	International Electrotechnical Commission
IEC JTC	joint technical committee of the International Organization for Standardization
IEEE	Institute of Electrical and Electronics Engineers
IEM	in-ear monitor
IESG	Internet Engineering Steering Group
IETF	Internet Engineering Task Force
IFPI	International Federation of the Phonographic Industry
IGF	Internet Governance Forum
IGO	International Governmental Organization
IMAP	Internet Message Access Protocol
IMT	International Mobile Telecommunications
InkML	Ink Markup Language
IoT	Internet of Things
IP	Internet Protocol address
IPA	International Publishers Association
IPFIX	IP Flow Information Export
IPPG	Internet Protocol Performance Metrics
IPR	Intellectual Property Rights
IRI	Internationalized Resource Identifier
IRIA	Institut de Recherche en Informatique et en Automatique
IRPC	Internet Rights and Principles Coalition
IRTF	Internet Research Task Force
ISAN	International Standard Audiovisual Number
ISDN	Integrated Services Digital Network
ISM	industrial, scientific, and medical
ISO	International Organization for Standardization
ISOC	Internet Society
ISP	Internet service provider
ISWC	International Standard Musical Work Code
ITA	Information Technology Agreement
ITAR	Traffic in Arms Regulations

ITU	International Telecommunication Union
ITU-R	International Telecommunication Union-Radio communications sector
ITU-T	Telecommunication Standardization Sector (a division of the ITU)
IPR	Intellectual property right
JEDEC	Joint Electron Device Engineering Council
JSON	JavaScript Object Notation
KSKF	Key Signing Key Futures
LAA	License Assisted Access
LAN	Local area network
LBT	Listen before talk
LDP	local differential privacy
LoA	Letter of Assurance
LoRa	Long Range
LPWAN	low-power wide-area network
LTE	Long Term Evolution
M2M	Machine-to-machine
M3AAWG	Messaging, Malware, and Mobile Anti-Abuse Working Group
MAC	Media Access Control
MIKEY-SAKKE	Multimedia Internet KEYing—Sakai Kasahara Key Exchange
MIME	Multipurpose Internet Mail Extensions
MIMO	multiple-input and multiple-output
MNO	mobile network operators
MMOG	mass online multiplayer game
MPAA	Motion Picture Association
MPC	multi-party computation
MPEG	Moving Picture Experts Group
MS	Multiple Streams
MU	multi-user
NAESB	North American Energy Standards Board
NATO	North Atlantic Treaty Organisation
NCRA	National Cooperative Research Act (1984)
NCRPA	National Cooperative Research and Production Act (1993)
NesCom	New Standards Committee (IEEE)
NGO	Non-governmental Organization
NIA	Network Advertising Initiative
NIPA	National IT Industry Promotion Agency (South Korea)
NIST	National Institute of Standards and Technology
NMI	Netmundial Initiative
NomCom	Nominations Committee
NSA	National Security Agency
NSF	National Science Foundation
NSS	National Standards Strategy
NTIA	National Telecommunications and Information Administration
OAA	Open Automotive Alliance

OASIS	Organization for the Advancement of Structured Information Standards
OCF	Open Connectivity Foundation
ODRL	Open Digital Rights Language
OECD	Organisation for Economic Co-operation and Development
Ofcom	Office of Communications (UK)
OFDMA	Orthogonal frequency-division multiple access
OMA	Open Mobile Alliance
OMB	Office of Management and Budget
ONR	Office of Naval Research
ORG	Open Rights Group
OSCE	Organization for Security and Co-operation in Europe
OSI	Open Source Initiative
OTA	Online Trust Alliance
OTT	Over-The-Top
OWL	Web Ontology Language
P2P	Peer-to-Peer
P3P	Platform for Privacy Preferences
PAE	Patent Assertion Entity
PAG	Patent Advisory Group
PAMW	PRISM Aggregator Message for Web Content
PAR	Project Authorization Request (IEEE)
PatCom	Patents Committee (IEEE)
PAW	Predictable and Available Wireless
PCI DSS	PCI Data Security Standard
PCI SSC	Payment Card Industry Security Standards Council
PD	Preamble Detection
PEARG	Privacy Enhancements and Assessments Research Group
PETS	Privacy Interest Group and Privacy Enhancing Technologies Symposium
PGP	Pretty Good Privacy
PHY	physical layer
PII	personally identifiable information
PING	Privacy Interest Group
PIPA	Preventing Real Online Threats to Economic Creativity and Theft of Intellectual Property Act
PIR	Public Interest Registry
PLMRS	Private Land Mobile stations
PLS	Pronunciation Lexicon Specification
PMSE	Programming Making and Special Events
PNG	Portable Network Graphics
POISED	Process for Organization of Internet Standards
POP	Post Office Protocol
PRIVSEC	Privacy and Security
PS	Public Safety

PSIG	Patent and Standards Interest Group
PTAB	Patent Trial and Appeal Board
PURL	persistent URL identifiers
QAM	quadrature amplitude modulation
QUIC	Quick UDP Internet Connections
RATS	Remote ATtestation ProcedureS
RDF	Resource Description Framework
REFIT	regulatory fitness and performance
REL	Rights expression languages
RFC	Request for Comments
RIAA	Recording Industry Association of America
RSA	Rivest–Shamir–Adleman algorithm
RSOC	RFC Series Oversight Committee
RTP	real-time transport protocol
R&TTE	Radio Equipment and Telecommunications Terminal Equipment and the Mutual Recognition of their Conformity
RTT	round trip times
SAC	Standardization Administration of the People's Republic of China
SAML	Security Assertion Markup Language
SCXML	State Chart eXtensible Markup Language
SDL	Services Description Language
SDMI	Secure Digital Music Initiative
SDO	Standard-developing organization
SDOAA	Standards Development Organization Advancement Act
SDP	Session Description Protocol
SEPs	Standard Essential Patents
SFD	specification framework document
SGML	Standard Generalized Markup Language
S-HTTP	Secure HTTP
SIG	Special Interest Group
SIP	Session Initiation Protocol
SISR	Semantic Interpretation for Speech Recognition
SKOS	Simple Knowledge Organization System
SMART	Stopping Malware and Researching Threats
SML	Standard Mark-up Language
SMTP	Simple Mail Transfer Protocol
SMPTE	Society of Motion Picture and Television Engineers
SOAP	Simple Object Access Protocol
SOLID	SOcial LInked Data
SOPA	Stop Online Piracy Act
SOUPS	Usable Privacy and Security
SNA	System Network Architecture (IBM)
SPARQL	SPARQL Protocol and RDF Query Language (pronounced 'sparkle')
SPF	Sender Policy Framework
SRGS	Speech Recognition Grammar Specification

SSL	Secure Sockets Layer
TAG	Technical Advisory Group
TANs	Transnational advocacy networks
TBT	Technical Barriers to Trade
TCP	Transmission Control Protocol
TCP	Transport Control Protocol
TETRA	Terrestrial Trunked Radio
TLD	top level domain
TLS	Transport Layer Security
TPE	Tracking Preference Expression
TPWG	Tracking Protection Working Group
T2T	Thing-to-Thing
TRIPS	Trade-Related Aspects of Intellectual Property Rights
TSDSI	Telecommunications Standards Development Society, India
TVWS	Television White Spaces
UDHR	Universal Declaration of Human Rights
UDP	User Datagram Protocol
UHF	Ultra High Frequency
UMID	Unique Material Identifier
UNESCO	United Nations Educational, Scientific and Cultural Organization
URI	Uniform Resource Identifier
URL	Uniform Resource Locator
UN	United Nations
UNECE	United Nations Economic Commission for Europe
UNEP	United Nations Environment Programme
URI	Uniform Resource Identifier
URL	Uniform Resource Locator
USB	Universal Serial Bus
USPTO	United States Patent and Trademark Office
VITA	Versa Module Europa (VME)bus International Trade Association
VLV	Voice of Listener & Viewer
VoIP	Voice over Internet Protocol
VPN	Virtual Private Network
W3C	World Wide Web Consortium
WAI	Web Accessibility Initiative
WAN	World Association of Newspapers
WCAG	Web Content Accessibility Guidelines
WEF	World Economic Forum
WG	working group
WGIG	Working Group on Internet Governance
WHATWG	Web Hypertext Application Technology Working Group
WHOIS	Who is responsible for this domain name?
WIPO	World Intellectual Property Organization
WLAN	wireless LAN
WPF	World Privacy Forum

WSA	White Space Alliance
WSDL	Web Services Description Language
WS-I	Web Services Interoperability
WSIS	UN World Summit on the Information Society
WTO	World Trade Organization
WWW	World Wide Web
WSDL	Web Services Description Language
XACML	eXtensible Access Control Markup Language
XML	eXtensible Markup Language
XrML	eXtensible Rights Markup Language

1

Global Standard-setting in Internet Governance

To the surprise of many in the industry, Microsoft adopted Mozilla's 'Do Not Track' (DNT) open standard for its Internet Explorer browser in 2011. Apple's Safari, Opera, and Google's Chrome followed suit. The feature communicates to websites that a user does not wish to be tracked and advertisers are expected to respect this choice on a self-regulatory basis. DNT opt-outs stem the amassing of expansive user information by advertising companies. The standard for DNT was agreed upon in the World Wide Web Consortium (W3C), a standards-developing organization (SDO) based at MIT in Boston. Promotion of the standard was led by an alliance of civil society groups lobbying for increased privacy for citizens in Internet use. DNT was subsequently backed by the US Federal Trade Commission (FTC), the Obama administration, and the European Commission. The choice to switch between 'opt-in' and 'opt-out' has a far-reaching impact on the ability of the advertising industry to collect and collate data on citizens. Although highly technical, such standards shape the behaviour of groups across the entire industry. This represents not an end but the beginning of a long-raging battle between different actors, companies, user groups, and states over data protection and privacy.

The SDO role in Internet governance is notable given the Internet's central place in society and relevance for economic growth. However, the way in which the Internet is managed is vastly under-studied partially due to how Internet governance has been framed by researchers. Most scholars working on Internet governance have focused on the role of international organizations such as the UN's Internet Governance Forum (IGF) and International Telecommunication Union (ITU), or the Internet Corporation for Assigned Names and Numbers (ICANN) which deals with Domain Name System (DNS) registration (Drissel, 2006; Malcolm, 2008; Slavka, 2008; Brown and Marsden, 2009; Christou and Simpson, 2006, 2007, 2009; Mathiason, 2008; DeNardis, 2009, 2010, 2014; Mueller, 2010; Froomkin, 2013; Glen, 2014; Scholte, 2017).

Some researchers have argued that the state still plays a pivotal role within these organizations (Goldsmith and Wu, 2006; Mueller, 2010; Marsden, 2012;

Global Standard-Setting in Internet Governance. Alison Harcourt, George Christou and Seamus Simpson, Oxford University Press (2020).
DOI: 10.1093/law-oso/9780198841524.001.0001

Brown, 2013; Nye, 2014). Mueller's work on 'networked governance' emphasizes the Internet's dependence on traditional state actors (2010:6). Goldsmith and Wu assert that ultimately the Internet is governed by states, given differences in implementation determined by geography and the mechanism of government coercion (2006). They argue that the United States, European Union, and China dominate Internet governance and that 'territorial governmental coercion is fatal to globalization theory as understood today, and central to understanding the future of the Internet' (2008:184). As Mueller and colleagues assert: 'if there is a single root, someone must stand at the top of the hierarchy' (2013:104).

Another body of literature points to the rise of Internet activism globally. Nowhere is transnationalism more evident than in Internet governance where rules and jurisdictions have no clear state boundaries. At the same time, 'the rise of a network society' (Castells, 1996) has enabled cross-border activism with digital advocacy groups creating transnational networks. Examples are the success of activist groups in preventing the US Congress from adopting the Stop Online Piracy Act (SOPA) and PROTECT IP Act (PIPA) as well as limiting the Real ID Act in the United States. They were also successful in stemming key revisions of the World Intellectual Property Organization (WIPO) Copyright Treaty and the Digital Millennium Copyright Act (DMCA). The SOPA and PIPA campaigns were headed by the Electronic Frontier Foundation (EFF), the Free Software Foundation (FSF), the Open Source Initiative (OSI), Publicknowledge, Fightforthefuture, and Demandprogress, and were supported by hundreds of regional and national organizations across the globe. As Lee explains, when 'EquanimousMind linked to the "veto SOPA" petition on Reddit... [t]he two petitions against SOPA elicited over 100,000 signatures in record time' (2013:75).

Activism was also key to the prevention of the EU's adoption of the Anti-Counterfeiting Trade Agreement (ACTA) which sought to establish a global legal agreement for preventing copyright infringement on the Internet. Horten's work details how digital rights activists convinced the European Parliament to reject ACTA, with a 478 to 39 vote against (2011; 2013). Key campaign groups within Europe included Against ACTA, Business Europe, Digitale Gesellschaft (Germany), Doctors Without Borders, European Digital Rights initiative (EDRi), La Quadrature due Net (France), MARQUES, the Open Rights Group (ORG), Pro ACTA, Panoptykon Foundation (Poland), and STOP ACTA (Horten, 2011; Dür and Mateo, 2014:1206). Dür and Mateo report 58 street protests against ACTA with 100,000 people taking part in Germany alone (2014:1207).

These European groups were supported by US organizations, specifically the American Civil Liberties Union (ACLU), Avaaz, the Software Alliance (BSA), EFF, FSF, and Lumen (formerly Chilling Effects). Avaaz collected 3 million signatures in an online petition against ACTA from across the globe (Dür and Mateo, 2014:1200). Activists not only came from different states but also from different contexts. The ACLU, Doctors Without Borders, and Avaaz advocated individual rights and liberties. The BSA and FSF campaigned for software freedom, whereas groups such as EDRi, La Quadrature due Net, and the Open Rights Group defended digital rights. STOP ACTA and Against ACTA were ad hoc groups formed for a sole purpose. Similar information groups were established to protest against the Hadopi law enacted to suspend Internet access for copyright infringement in France in 2013 (Brown and Marsden, 2013; Dür and Mateo, 2014). Hadopi was revised when it was demonstrated that it was ineffective and costly. These and other cross-national campaigns have led to groups linking up across borders and the formation of global Internet activist groups such as the Internet Social Forum,[1] established in 2015 at the World Social Forum as a 'Call to Occupy the Internet'.[2]

A further large body of literature focuses on the trend towards having multiple stakeholders in Internet governance which is seen as a mechanism for 'bridging the gap' between international and national legal systems (Malcolm, 2008:26). A multi-stakeholder agenda in Internet governance has been promoted since the 2003 United Nations World Summit on the Information Society (WSIS), which introduced civil society inclusion. Important partners in Internet governance are the IGF, the Global Knowledge Partnership (GKP),[3] and NETMundial, which act as multi-stakeholder fora for defining major issues. However, these multi-stakeholder groups have substantial limitations as they have no decision-making power (Drake and Wilson, 2008; Mueller, 2010). The IGF's main success was to provide pressure to move ICANN away from ultimate control by the US government resulting in the Internet Assigned Numbers Authority (IANA) transition.

The bulk of Internet governance does not take place within international organizations but in self-regulatory fora, such as the Internet Engineering Task Force (IETF) and the W3C, which have no state or public-sector body membership. These fora develop standards for interoperability between browsers, devices, and network technologies. Agreements are immensely technical and

[1] http://internetsocialforum.net/isf/.
[2] http://internetsocialforum.net/isf/?page_id=832.
[3] http://www.gkpfoundation.org/.

largely inaccessible to the non-expert. Accordingly, Internet governance similarly remains immensely technical and largely inaccessible to the lay person notwithstanding the fact that the Internet's role in economic development and the public sphere is increasingly central. The main members of SDOs in the field of Internet governance are based in the United States and Western Europe.

There is a growing literature on Internet standard-setting and protocol governance which this book builds upon (Bygrave and Bing, 2009; Farrell, 2009; Gencer, 2012; Brown and Marsden, 2013; DeNardis, 2014; Simcoe, 2014). This book aims to situate these debates in the wider field of standard-setting governance (Braithwaite and Drahos, 2000; Hall and Biersteker, 2002; Drezner, 2004; O'Neill and Van Deveer, 2004, Farrell, 2006; Jupille, Mattli, and Snidal, 2013). This body of literature sees standards' fora as dominated by corporate actors with little role for the state and virtually no role for civil society.

The Internet standards literature to date has focused on organizational aspects of SDOs (Bygrave and Bing, 2009; Farrell, 2009; Gencer, 2012; Brown, 2013; DeNardis, 2014; Simcoe, 2014). Most empirical studies focus on DNS, IPv6, and the initial history of TCP/IP or wider legal structures for Internet governance (Bygrave and Bing, 2009; Brown and Marsden, 2013; DeNardis, 2014; Ziewitz and Pentzold, 2014; Hofmann, Katzenbach, and Gollatz, 2017). DeNardis defines Internet standards as 'the agreed upon rules for providing specifications for formatting, encoding, compressing, error checking, encrypting and otherwise exchanging information electronically' (2015:72). Even though decision-making within fora is not visible, as DeNardis points out, the standards themselves, such as WiFi (Wireless Fidelity), MP3, HTTP (HyperText Transfer Protocol), VoIP (Voice Over Internet Protocol), and TCP/IP (Transmission Control Protocol/Internet Protocol) are now familiar enough to have become household terms.

Some authors have stressed the importance of promoting human rights and the public interest in protocol development (DeNardis, 2014, 2015; Cath and Floridi, 2016; Ten Oever, 2018, Ten Oever and Beraldo, 2018). Cath and Floridi claim that technologists have the 'responsibility . . . to protect human rights-by-design' in standard-setting. They adopt the term from Sweet and Schneier's work on responsibility by design which entails social responsibility being taken into consideration in the impact of technology.

Other authors have concentrated on the role of large corporations and market developments (Alverstand and Lie, 2009; Simcoe, 2014; Greenstein, 2017; Ten Oever and Beraldo, 2018). Ten Oever and Beraldo claim there is an 'ossification' of the Internet which consists of the presence of monopolies, oligopolies, or gatekeepers, and a decrease in transparency. They point not only to the more visible market players such as Amazon, Google, Facebook, Apple, and Microsoft but also leaders in content distribution (Akamai and Cloudflare), telecommunication providers (China Mobile, Verizon, AT&T, and Vodafone), networking hardware (Cisco and Huawei), and chip manufacturing (Intel and AMD). This literature suggests that there is market consolidation within various points of the Internet stack and layer consolidation particularly by content delivery systems such as AWS.

However, neither the literature on global Internet activism nor governance tells us much about the decision-making processes and interactions of actors *within* SDOs as yet. Specifically, how is the public interest promoted? Are consumer and digital rights groups able to defend citizen rights against corporate interests? Have corporate goals been balanced by the public interest? The book addresses this gap by investigating within-case (Brady and Collier, 2010) analyses of decision-making leading to specific standards. It utilizes Scholte's definition of civil society which is understood to be 'a political arena where associations of citizens seek, from outside political parties, to shape the rules that govern one or other aspect of their common life' (2007:311).

Standards are important for interoperability and to prevent an individual or an organization being locked in to a single dominant (particularly commercial) entity. Standards-Developing Organizations (SDOs) enable a wider technical community to scrutinize proposals for possible errors and security. Nonetheless, procedures for decision-making are complex and often opaque and core decision-making often takes place outside the public gaze. As DeNardis observes, the process of setting standards for the Internet raises 'difficult questions about what interests shape these standards; how the public interest is reflected in design decisions; and what sources of institutional legitimacy are necessary for largely private institutions to perform policymaking function' (2014:65–6). At the same time, standard development is crucial to the way in which citizens utilize the Internet. There is a need to understand standards contextually in terms of organizational identity and culture (Alberstrand and Lee, 2009:127; Gencher, 2012). A key component of this is the nature and roles played by SDO memberships (Kanevskaia and Zingales, 2018).

Recent years have seen a high degree of spill-over of politicized policy areas such as digital rights management, data protection, security and spectrum to SDOs, policies which were formerly domains of the nation state. This spill-over has been triggered by key events such as when the Central Intelligence Agency (CIA) whistle-blower, Edward Snowden, leaked highly classified information from the National Security Agency (NSA) to journalists at the *Guardian, Washington Post, New York Times* newspapers, and *Der Spiegel* magazine.

Snowden's revelations that the US NSA and UK Government Communications Headquarters (GCHQ) were capturing vast amounts of Transport Layer Security (TLS) traffic greatly increased SDO work on privacy and securing network traffic, recently culminating in, amongst other work, the adoption of the TLS 1.3 standard by the IETF in March 2018. Decision-making is technologically based but intrinsically it is political. These newer developments are a focus of this book.

Despite backlash from the engineering community over Snowden's actions, companies are facing increasing pressure from governments and other stakeholders for the provision of personal data and the enabling of surveillance technologies. In 2018, part of the US Foreign Intelligence Surveillance Act (FISA) Amendments Act enabling government surveillance programmes was renewed. This authorizes US security agency requests made to Internet service providers and technology companies through the Foreign Intelligence Surveillance Court.

Social media have also become targets for government surveillance. Even status codes have become politicized with the IETF developing the code 451 to signal government blocking. Indeed, security questions abound with the recent explosion in the Internet of Things. In parallel, digital rights groups and consumers are pressing the same companies for increased emphasis on consumer and client privacy and consent as demonstrated by public reaction to the Cambridge Analytica scandal in 2018 when it emerged that the company had mined the personal data of millions of Facebook users without their consent for political advertising purposes.

In addition to security, SDOs are grappling with the efficiency of cloud storage, limits of spectrum use, and autonomy and management of devices. Despite this increasing politicization, the importance of SDOs for ensuring and maintaining the openness, interconnectivity, and security of the Internet are crucial. Standards are important for interoperability and to prevent lock-in to single company's technology. SDOs enable a wider technical community to scrutinize proposals for errors and security. Nonetheless, decision-making operates outside the public gaze even though standard development is crucial to the way in which citizens utilize the Internet.

There is some recognition and movement towards the inclusion of public interest goals within standard fora. In 2012, a Joint Statement of Affirmation was signed by the IETF, the Internet Society (ISOC), Internet Architecture Board (IAB), W3C, and Institute of Electrical and Electronics Engineers (IEEE), which commits these organizations to an 'open participatory process.'[4]

[4] https://open-stand.org/about-us/principles/

The W3C has introduced horizontal measures on security, web accessibility, and internationalization which must be considered in all W3C working groups.

In 2014, the IETF developed a policy on addressing pervasive monitoring with technical solutions (IETF, 2014). In 2015, it also initiated a Human Rights Protocol Considerations Research Group (HRPC) to assess the ability of IETF standards to 'enable, strengthen or threaten' human rights. Meanwhile, in the same year, the IEEE commenced its Ethics Program which introduces professional guidelines and codes of ethics into IEEE work. Implementation thus far has taken place within the IEEE Global Initiative on Ethics of Autonomous and Intelligent Systems[5] and the Ad Hoc committee on Design for Ethics, and events and publications which incorporate 'human values' in the design of complex systems.

The following chapters will explore the development of these and other measures and the extent to which the public interest has been represented in standard-setting to date. The book explores the interplay between different interests within SDO fora and whether civil society and other actors are able to defend and promote citizens' rights. In terms of theoretical lenses on decision-making processes, the book extends and applies the Multiple Streams (MS) approach for analysis of SDO decision-making through illuminating its explanatory utility in cases of governance—but not government—at the international level. Each chapter delves into the working practices of SDOs through following the progression of protocol development.

The book's case studies were selected via a two-step process. First, a survey which ran from 2016 to 2017 was conducted as a part of the ESRC International Professional Fora: A Study of Civil Society Organisation Participation in Internet Governance project. Two hundred and twenty-four detailed responses were received chiefly from participants in the W3C, Oasis, IAB, IETF, ISOC, the IEEE, and IRTF. This was assessed to be a representative sample based on an estimated proportion of the number of overall active participants in fora identified in Chapter 3. The survey enabled the project team to identify the key political problems under discussion in SDO fora and understand how fora are targeted by civil society actors. In the second step, 300 semi-structured interviews were conducted between 2015 and 2019 with SDO participants, civil society and digital rights group representatives, civil servants, and government officials. Face to face semi-structured interviews were conducted in Berlin, Boston, Brussels, Chicago, London, Milan Mountain View, Prague, San Francisco, San Diego, Tallinn, and others via Skype where this was

[5] https://standards.ieee.org/industry-connections/ec/autonomous-systems.html?.

not possible. As a part of the project a number of workshops were held with SDO participants, policy experts, and relevant civil society sector representatives. The workshops aimed at building links with experts and practitioners involved in Internet governance and to discuss how such actors defined their positions in the case study areas. The focus was on public interest solutions to policy problems and exploring connections across the participant organizations, particularly between SDO participants and third-sector interests. The workshops and interviews provided information on SDO decision-making processes, lobbying tactics, and different stakeholder positions from all sides of the decision-making process. Additional methodology included the analysis of mailing lists, official SDO publications, statements by different advocacy groups, and attendance at SDO meetings.

Outline of the Book

Chapter 2 outlines the theoretical grounding on which the book is premised. This consists of two elements. The global governance literature which facilitates an explanation of which actors can affect policy in international self-regulatory fora. The second relates to the MS framework which enables us to examine solutions proposed by different actors within these fora. This facilitates the investigation of different solutions which are framed within 'policy streams'. The framework enables the identification of actors and how they couple solutions, and why, and under what conditions they are most effective. In doing so, a series of mechanisms are identified which enable or constrain decision-making. Through these twin lenses, actors representing the public interest can be identified and answers offered as to whether solutions address public interest concerns.

Chapter 3 provides an overview of the internal organization and main standard areas of Internet-focused SDOs. It explains the four main layers (Application, Transport, Internet, Network Access) of the Internet stack and focuses on the principal SDOs associated with them: W3C, OASIS, the IETF, and IEEE. It analyses SDO governance structures and their most salient areas of work. The central purpose of the chapter is to provide a governance context to understand the decision-making process in each of the standards investigated in the case study chapters. It plots the evolution of each SDO, its main purpose, functions, and central decision-making processes, highlighting the differences, as well as the degree of synergy and collaboration, between each. Procedures differ between the fora and moving goalposts and high barriers to entry make it difficult for civil society to participate. Chapter 3 describes which areas of work are concentrated in which SDO fora and why, with examples of forum-shopping by companies.

Problems of security, capacity, the pace of technological development, internationalization, and Intellectual Property Rights (IPR) cut across all ongoing SDO work. Security, particularly pervasive monitoring, became a core issue for SDOs post-Snowden. The number of working groups within SDOs dealing with aspects of security issues has increased dramatically in recent years. At the same time, the speed of technical developments, such as continuous release of ever-green browsers coupled with the demands for integrated features and interoperability for the Internet of Things (IoT), has fragmented decision-making. IPR too has always been contentious within SDOs but has gained in salience due to the increase in the need for Standard Essential Patents (SEPs) and the convergence of sectors. We are now moving into a world where every bit of digital data is labelled and formally granted permission not only for personal use but for trade in digital goods between businesses. For this reason, the case study chapters focus on developments affecting privacy/security, IPR and copyright, and mobile communications standards respectively. Chapters 4–7 focus on internal and external mechanisms for policy change in private sector governance. Chapters 8–11 explain how the external environment, state actors, and key events have influenced SDO decision-making.

Chapter 4 examines the activities and impact of Google on IETF security protocols. For twenty years, Secure Sockets Layer (SSL) and its successor, Transport Layer Security (TLS), provided security for Internet traffic. However, the TLS 1.2 protocol, developed in 2008, suffered from a series of implementation and security issues. Push-back from companies to demands from national security agencies for access manifested itself in SDO fora post-Snowden. In 2016, Cisco, Fortinet, and Juniper revealed that the NSA had successfully targeted its firewalls for years. However, stasis within the IETF barred upgrade to TLS 1.3. The chapter analyses the emergence of and controversies around potential solutions to TLS and how the parallel development of the Quick UDP Internet Connections (QUIC) protocol by Google opened a window of opportunity to enhance security through the development of both QUIC and TLS 1.3 within the IETF. It explains how Google's QUIC standard created an opportunity not only to address the issue of latency but also enhanced security (encryption) through ongoing work on TLS. This resulted in agreement on TLS 1.3 in March 2018. The overall consequences of Snowden on IETF work is a key theme of the chapter, with the parallel development of the Pervasive Monitoring RFC7258. The latter sees any monitoring as a 'technical attack that should be mitigated where possible through the design of protocols'. This has changed the overall ideational environment in which engineers now work within the IETF.

The W3C is a recognized standards organization with formalized consensus-driven rules and membership fees. As a result, its progress and agreement on key standards for the Internet can be slow due to a large degree of scrutiny by many members. The W3C began working on HTML5.2 in 2014 which, although widely implemented, remained at the proposal stage until it was formally adopted in December 2017. By contrast, the Web Hypertext Application Technology Working Group's (WHATWG) Living Standard for HTML is a fast-moving specification, which is continually updated, mainly developed between a few US West Coast-based browser companies dominant in the market, Apple, Google, Microsoft, and Mozilla. This close cooperation between the major browsers was consolidated in 2017 when Microsoft formally joined WHATWG after it revised its IPR policy.

Chapter 5 details how parallel development of the HTML standard between the two standards fora has created shifts in the focus of W3C work and membership. These internal shifts have opened opportunities for the transfer of a number of new areas of work within the W3C, most notably in the development of web devices and television standards However, as standards are voluntary and agreement is reached on a self-regulatory basis, protocol development continues to be shaped by external legal environments. In this context, the second part of the chapter explains how the platform for DRM moved to the W3C.

Development of EME was shaped by the US DMCA and 2019 EU Directive on Copyright in the Digital Single Market which mandate that online platforms use technical measures to prevent copyright infringement. The drive by states for technology solutions has brought about a shift in copyright protection away from international treaty-based agreements to technical standard agreements in fora such as the W3C. This shift has rebalanced the power of decision-making in favour of rights holders in the global copyright debate away from digital rights activists as the debate is reframed in technical terms.

Chapter 6 turns to standards developed for mobile communications in the IEEE. The local area network wireless environment has been the setting for the co-existence of licensed mobile communications operators and unlicensed WiFi Internet access providers historically. The IEEE802.11 family of standards, developed for WiFi services, has dominated standards-setting in this space.

Recent demand for new services, particularly from the growth in licensed mobile phone use but also from IoT and machine-to-machine (M2M) communication, has led to intense competition for free to enter unlicensed capacity and thus made co-existence uneasy. The chapter focuses on the IEEE's efforts to create the 802.11ax standard for WiFi and the challenges faced from the internal and external standards making environments. It shows how 802.11ax development was highly controversial within the IEEE itself, with concerns about special interest group (SIG) dominance in the development process becoming a key issue. It also illustrates how alternative standards making contexts to the IEEE, based on licensed spectrum standards traditions, aimed to develop standards to exploit unlicensed spectrum ahead of the IEEE initiative, asserting co-existence friendly claims, in the process. The chapter concludes by evaluating the unlicensed spectrum standards co-existence debate and considers the likely impact of the 802.11ax standard as a policy solution.

In Chapter 7, the development of DNT in the W3C is analysed as a response to stagnation and policy drift in the standard-setting environment. Tracking behaviour has sought to ensure that users have a choice regarding whether they are tracked or not online. The proliferation and persistence of online tracking of consumer behaviour through multiple technologies for the purpose of collecting information and constructing targeted ads based on the profiling of the interests and preferences of users, at surface level, seems relatively harmless. From a marketing, advertising, and publishing perspective, such activity is crucial to providing online content and ultimately revenue (and profit) generation . However, the broader impact and implications of such activity is much more wide-ranging, in particular given that much of it is not transparent and that it can result in the violation of a plethora of individual rights.

The chapter focuses on the extent to which different interests and stakeholders converged on the conception of the DNT standard and its development with specific focus on the role of civil society. The case is explored in the broader (self-) regulatory, legal, and political environment within which the DNT standard has been developed, agreed, and implemented. To this end, the chapter analyses the appearance of the problem, how a consensus (Candidate Recommendation) was arrived at in the W3C, missed windows of opportunity that arose in the development process, and the extent to which the DNT standard has been adopted and enforced. It focuses, in particular, on how the evolving policy and political streams in the European Union and United States relating to privacy have exerted pressure on the relevant actors (in particular on ad networks) to honour DNT.

Chapter 8 turns to the role of external factors on the products of SDO in the field of radio spectrum, specifically television white spaces (TVWS). These are spectrum buffer zones originally allocated to prevent interference between broadcast signals. Historically, these spaces were occupied by providers of Programme Making and Special Events (PMSE) services deployed in large venues and productions. However, increasing demand for spectrum led to pressure for TVWS use to be made more extensive through opening it up to a wider range of users and lower power Internet of Things (IoT) applications. Initiatives to reform TVWS use are underpinned by vital technical standards with the aim of introducing computer-mediated Dynamic Spectrum Access (DSA) to implement interference free use of the wireless environment. The chapter shows how technical standards-making for TVWS development creates outcomes which are heavily contingent on external actor preferences in the spectrum policy making domain. The cases of TVWS policy in the USA and the UK explored in the chapter show evidence of strong contestation from incumbent interests in commercial, public service, and civil society quarters to policy change. They also illustrate the need for effective market development to exploit the affordances technical standards development.

Following on from Chapter 4 which assessed TLS development, Chapter 9 analyses push-back from the standards community against pervasive monitoring by the state through examining other standards targeted for state surveillance, with three case studies. The first part of the chapter investigates implementation of the IETF's policy on pervasive monitoring. Following Snowden's activities, the IETF has worked on identifying protocols subject to security agency attacks and eavesdropping.

The chapter then charts advocacy efforts of digital rights groups first in the work of the HRPC in the development of protocol 451, which signals Internet blocking by governments. The HTTP status code 451 'unavailable for legal reasons' was adopted by the IETF in December 2015. Internet service providers and browsers often receive requests to take down content. Normally in these cases, websites would state the 403 ('Forbidden') or 404 ('Page Not Found') error code. The 451 code now shows when a website has been blocked by ISPs due to a government or court request or legal order. The idea for the protocol came from a blog post by Terence Eden at the UK Open Rights Group. Chapter 9 details the introduction and build-up of support within the IETF that led to the code's adoption in 2017. The latter part of the chapter explores the actions of states through the development of government positions in the United Kingdom and United States, most notably against the company Huawei. It analyses the position of the engineering community within these contexts.

Chapter 10 continues with less visible work on protocol development which nevertheless substantially affects users. The first part of the chapter describes work to mitigate browser fingerprinting within the IETF and W3C. Devices such as computers, cameras, and mobile phones can be identified by remote observers based upon device users' activity. Big data companies sell software which 'fingerprint' web browsers and analyse use without users' knowledge even when these users are aware enough to block browsing cookies and use virtual private networks (VPNs). The chapter details how public campaigns advanced by the EFF led to the introduction of a W3C Candidate Recommendation in April 2016 and an IETF RFC7230 in 2014 which mitigate the identification of user devices. It provides evidence that engineers are able to forward public interest initiatives in conjunction with academics, digital rights groups, and wider civil society organizations within SDO fora.

Chapter 11 continues in its assessment of the role of the state as an external actor. The main participants in SDOs in the field of Internet governance are based across the United States and Western Europe. However, discrepancies between US and European intellectual property right (IPR) laws have created forum-shopping (Weston and Kretschmer, 2012). The United States recognizes the patentability of software whereas the EU restricts patents on software, formulae, and algorithms. For this reason, most patents in the high-tech industry are registered in the United States. However, given the rapid development of the IoT, where an estimated 20 billion devices will be connected to the Internet by 2020 (Hung, 2017), standard essential patents (SEPs) are on the rise. The chapter shows how this has led to convergence between the United States and the European Union on IPR policies. This has resulted in coordinated pressure on SDOs to reform their SEP policies. This occurred gradually through, initially, US and EU competition decisions and legislative changes leading to direct statements in the form of US Department of Justice Business Review letters and workshops and opinions expressed by the European Commission.

Chapter 12 presents conclusions about the state of play in SDO fora with significance for the wider literature on global governance. It analyses findings of the empirical cases on a comparative basis in order to explain what triggers policy adoption. Utilizing the lens of the MS framework, it identifies and explains the processes under which SDO rules of interaction are established at the international level and whose interests are represented in SDO fora, and how. It explores the occurrence of policy entrepreneurship, forum-shopping, and coupling uncovered within the book's chapters. It illustrates how policy problems access the agenda, how and where constellations of actors carry out their search for solutions, and how the selection of options is biased. The chapter analyses how internal actors and external mechanisms have been able to steer agendas, resulting in policy change. Overall, it offers an explanation of the framing and coalition-building strategies of important actors, their ability to influence the agenda, and the mechanisms that lead to eventual decisions.

According to the global governance literature, one would expect corporate groups to lead professional fora with a steer from key member states, particularly the United States. However, the book's findings suggest a different and more nuanced pattern emerging, with increasing inclusion of public interest goals within SDO fora. The second part of the chapter describes how the public interest has been represented, drawing upon the case studies. The conclusion investigates areas in which civil society and digital rights groups could further seek to influence SDO fora, in particular in relation to ensuring privacy and transparency within protocol development, with an examination of legal structures at national and international levels. Overall, the book contributes to the academic debates on democracy and the Internet, global self-regulation and civil society, and international decision-making processes in unstructured environments.

2

Informal Governance and Decision-making Through Multiple Streams

Explaining Standard-developing Organizations

This chapter sets out the twin theoretical premises of the book which relate to global governance at the level of actor influence and the multiple streams (MS) framework at the level of decision-making within standards-developing organizations (SDOs). To this end, the first part of the chapter provides critical discussion of the literature on the influence of actors in transnational and international self-regulatory fora and demonstrates the necessity of further work to address the complex question of which actors uphold the public interest in SDO fora where states and public actors are largely absent.

The second part of the chapter demonstrates the case for adapting and utilizing the multiple streams (MS) approach in public policy to the analysis of the SDOs under investigation. The MS framework is utilized in order to provide a more in-depth and nuanced explanatory account of how problems appear on SDO agendas, the nature of the contestation relating to the discussion of standards development, and the extent to which and how standards are translated to final adoption. In doing this, the book identifies mechanisms at the conceptual level derived from the global governance literature. These facilitate the understanding of how deliberation of standards development lead to agreement and policy decisions. Hence, the chapter develops the analytical framework on which the book is premised.

Transnational Governance

Transnationalism has opened new spaces for societal actors within global governance (Cerny, 1995, 2010; Risse-Kappen, 1995; Strange, 1996; Scholte, 2000, 2017). Due to a significant rise in participation in global politics, civil society

Global Standard-Setting in Internet Governance. Alison Harcourt, George Christou and Seamus Simpson, Oxford University Press (2020). © Alison Harcourt, George Christou, & Seamus Simpson.
DOI: 10.1093/law-oso/9780198841524.001.0001

became an important area of study for international relations following the Second World War. These actors are seen to balance private sector influence by representing public interest concerns in global governance fora (Steffek, Kissling, and Nanz, 2008; Jönsson and Tallberg, 2010). Their presence adds legitimacy to unelected international institutions (Held, 1995; Schmitter, 2000; Scholte, 2002, Steffek, 2010).

The end of the Cold War signalled new developments at the global level with a move towards increased transnational commerce and private sector governance. Building on Keohane and Nye's original work on transnational relations and world politics (1998), there has been a growing literature documenting the rising phenomenon of transnationalism. Strange documented the 'retreat of the state' with delegation of powers to private authorities (1996). Conceptualization work on transnationalism has been mainly based on case studies. The Routledge Research in Transnationalism series published in the early 2000s demonstrated transnational arenas in a number of areas including diaspora communities such as Turkish and Kurdish communities in Germany, transnational ageing, transnational families, marriage, migration, identity, and language. The Routledge Studies in Global and Transnational Politics looks at newer experiences in transnational media, higher education, ideas of freedom and globalization and localization.

The literature on transnational advocacy networks (TANs) tells us that activists are gaining ground at the international level. Scholte points to the emergence of a global civil society (2000). This literature views activist groups as 'norms entrepreneurs' (Finnemore and Sikkink, 1998) promoting a 'justice cascade' (Sikkink, 2011). This literature has noted the rising power of these groups globally (Armstrong, Bello, Gilson, and Spini, 2010; Cerny, 2010; O'Brien, Goetz, Scholte, and Williams, 2000; Trondal, Marcussen, Larsson, Veggeland, 2012). Hall and Biersteker (2002), in particular, identify activism to be highly effective at the international level. In the domain of human rights, Civil society organizations (CSOs) have exercised autonomy from the preferences of governments. Slaughter documented how 'transgovernmental networks' are increasingly in operation (2004). Cerny's theory of 'transnational neopluralism' explains the link-up of transnational actors at the global level and the move away from domestic dominance (2010).

There is general agreement in the literature that civil society is largely representative of public interest goals. The literature distinguishes between non-governmental organizations (NGOs) and wider transnational social movements such as anti-globalization or environmentalism which have defined sets of opinions and beliefs. Many authors distinguish between what

DeMars terms 'advocacy NGOs', which challenge governments, and 'service NGOs', which work in partnership with governments (DeMars, 2005). DeMars points out that groups can be private or third-sector actors but emphasizes the importance for transnational NGOs to have a 'global moral compass', 'bind themselves to societal and political partners in several countries', and 'facilitate the transnational movement of norms, resources, political responsibility, and information' (DeMars, 2005:12). According to Tarrow, groups representing the public interest should respond to 'collective challenges (to elites, authorities, other groups or cultural codes) by people with common purposes and solidarity in sustained interactions with elites, opponents and authorities' (1994:18).

Scholte argues that civil society can go a long way to providing legitimacy to global governance. This can be via: education about fora; public debate of decision-making providing critical deliberation; connecting citizens to fora activities via consultation exercises; putting pressure on fora for increased transparency about their operation; serving as a watchdog via monitoring; countering social hierarchies which hinder access to certain citizen groups; and providing recognition for political identities which are marginalized in society (2007:314). Finally, civil society can enhance the legitimacy of international organizations and self-regulatory fora by promoting charismatic leadership, which Scholte has argued occurred in the case of Berners-Lee's leadership of the W3C. This legitimacy building comes with significant caveats, however (Scholte 2007:301–321). Scholte firstly reminds us that contributions from other legitimizing actors such as parliaments and the mass media as well as industry should not be discounted. Second, civil society representatives can sometimes erode the legitimacy of global governance (2007:313). Third, CSOs themselves are not always legitimate. Sometimes advocacy groups are not agreed on what they are advocating (2007:320). Finally, legitimacy is not an end goal of global governance.

Much of the literature on global Internet governance documents civil society participation in UN Internet governance fora and multilateral treaty agreements. The United Nations began to integrate non-governmental organizations from 1945. As Scholte points out, civil society was conducive to adoption of the San Francisco Charter which established the United Nations. Under Article 71, the United Nations 'may make suitable arrangements for consultation with non-governmental organizations which are concerned with matters within its competence'. Götz documents the history of NGO involvement within the United Nations culminating in the creation of the United Nations Environment Programme's (UNEP) 'Civil Society & NGOs Unit' followed by the 'Major Groups and Stakeholders Branch' in 2004 which is still in existence today.[1] He also notes a change in terminology in the late 1990s from 'NGOs' to 'civil society organizations' consultation. NGO fora flank UN conferences such as the G7 (Götz, 2008).

Studies on civil society in Internet governance focus on participation in the UN's Internet Governance Forum (IGF), the Internet Telecommunications Union (ITU), which is a UN specialized agency, or the UN's World Summit on the Information Society (WSIS) and its Working Group on Internet Governance (WGIG) (Drissel, 2006; Christou and Simpson, 2006, 2007, 2009; Mathiason, 2008; Mueller, 2010). Civil society in WSIS was instrumental in pushing for the Internet Assigned Numbers Authority (IANA) transition under the Internet Corporation for Assigned Names and Numbers (ICANN). The United Nations has a historically high level of civil society participation due to a 'participatory norm of governance' (Steffek, 2010:83; Jönsson and Tallberg, 2010; Tallberg, Sommerer, Squatrito, and Lunds, 2013). Scholte demonstrated how civil society has increased legitimacy of global governance institutions (2007), but what happens when in fora with low participatory norms of governance?

Existing literature has observed cases wherein participation is 'virtually non-existent' such as immigration policy, security (in the North Atlantic Treaty Organization (NATO) and the Organization for Security and Co-operation in Europe (OSCE)) and finance (in the International Monetary Fund (IMF) and the Bank for International Settlements (BIS)) (Jönsson and Tallberg, 2010:238–9). Institutions with low levels of participation include the IMF and the World Trade Organization (WTO) which are seen to 'pay lip service to the norm of participatory governance, as a means to silencing or limiting opposition' (2010:242). Variation in civil society access can be explained by differing opportunity structures that international organizations provide based upon resource demand (Steffek, 2010:69). Policy areas with higher access such as environmental and human rights protection have higher policy expertise, implementation, and monitoring needs than other areas such as security and finance, which explains varying degrees of openness. In addition to resource demand, entry points for civil society are highly dependent upon the global script of activist groups themselves as well as the level of autonomy and willingness of the leadership of international organizations to enable participation (Liese, 2010:106). Path dependence of international institutions can also explain the lack of change (Causla Vifell, 2010). Green (2010) and Kalm (2010)

[1] http://web.unep.org/about/majorgroups/.

observe that states are unwilling to involve civil society in areas where there are high sovereignty costs.

Most of the transnationalism literature focuses on international organizations where states have formalized roles. However, a significant amount of Internet governance takes place in standard-setting fora such as the World Wide Web Consortium (W3C), the Institute of Electrical and Electronics Engineers (IEEE), and the Internet Engineering Taskforce (IETF), which have no formal state or involvement. Decision-making in these fora is dominated by private actors, mainly large corporations. One strand of the literature on private authority looks specifically at standards-developing bodies. Studies have been conducted in commercial law (Cutler, 2003), telecommunications (Salter, 1999), trade (Loya and Boli, 1999), accounting (Mattli and Büthe, 2005), and food and financial services (Porter and Ronit, 2010). This body of literature documents a loss of sovereignty as large multinational corporations move to private sector governance at the international level thereby bypassing state scrutiny. As Stone and Ladi note, 'when we drop the "public", it is also in recognition that authority becomes more informal and privatized' (2015:840).

Case study research on SDOs shows that these fora have long been dominated by corporate actors (Braithwaite and Drahos, 2000; Drezner, 2004; Farrell, 2006; Hall and Biersteker, 2002; O'Neill and Van Deveer, 2004, Jupille, Mattli, and Snidal, 2013). It documents the erosion of state power and public sector actors due to globalization with a move towards self-regulated systems of governance by corporations (Braithwaite and Drahos, 2000; Picciotto, 2011). Authors point to the rise of self-regulation and the political power of global corporations, law firms and active-knowledge politics which dominate transnational legal orders (Delazay and Garth, 1998). It informs us that the power of large corporate actors undermines national democratic processes raising important questions of representation of the public interest at the international level.

When assessing decision-making in SDO fora, much of this literature focuses on internal bargaining between private actors and the exogenous role of the state. Different studies demonstrate how the size and importance of a state's market are key determinants of corporate actor behaviour within international fora (Drezner, 2004; Goldstein and Steinberg, 2010). Benkler (2006) and Farrell (2006) demonstrate how corporate actors were pressed into adopting standards advocated by states through 'points of control' such as the threat of blocking software as well as national standard-setting. External actors have been influential in SDOs by targeting key points in the supply chain for leverage

(Garcia-Johnson, 2000). Garcia-Johnson showed how the chemical industry agreed to certification by external actors as it unburdened individual firms.

Mattli and Büthe demonstrate how actor decision-making in SDOs is influenced by domestic (2013:14), institutional (2013:54–60), market (2013:129–212), and regional coordination mechanisms (2013:95–6). They argue, for example, that in certain policy areas actors in self-regulatory fora are affected by the inability of regional standard-setting organizations due to regional coordination mechanisms to speak with a 'single voice' at the international level (2011:100). Other studies show how the reputational effect of standards can lead to SDOs agreeing to external certification in production processes (Prakash and Potoski, 2006) and labour standards (Abbot and Snidal, 2009).

However, none of the studies on standards bodies have yet assessed the role of civil society participation in SDO decision-making. This book builds upon this body of literature by explaining what role civil society plays. Internet governance is an important case study because it is increasingly decided at the global level but it affects almost every citizen at national and local levels daily. The book contributes to the literature on global governance and transnational advocacy in specific ways. First, it assesses the scope for civil society participation measured against the potential points of access highlighted in the wider literature on transnationalism as outlined earlier. Second, it evaluates the effect of external activism on Internet fora developments through the lens of the literature on transnational advocacy.

The advocacy literature has so far been interested in studying the role of activists in international organizations with some form of state involvement. Empirical studies observe how activist groups have been incorporated into international fora thereby bridging the representation gap (O'Brien, et al., 2000; Scholte, 2007). O'Brien and colleagues (2000) examined the interaction between multilateral economic institutions (such as the IMF, the World Bank, and the WTO) and global social movements. They label this 'complex multilateralism', a form of incremental reform of international institutions to accommodate civil society input (O'Brien, et al., 2000; O'Brien, 2006). Rather than seeing the relationship as solely adversarial, they show a move towards active inclusion of activist groups. Although this literature is extensive, not much is known about role of global activists vis-a-vis SDO decision-making with no formal state involvement.

How well then is the public interest represented in the closed environment of Internet standard-setting? What points of access are available to civil society actors and transnational activists? The book's findings show that civil society is unable to translate complex technical issues effectively to wide target audiences.

Leverage is lost with the move to SDOs as there is no pivotal decision-making body. Civil society is limited to informal mechanisms for participation and a default fallback to external activist strategies. Indeed, gradual changes in the political architecture towards increasing use of self-regulation render current Internet infrastructure developments often immune to external campaigning tactics by civil society actors. These conclusions point to an overall erosion of participatory governance.

Whereas international organizations with state participation have support measures designed to engage citizens in public consultation of decision-making which affect them, SDOs restrain this access. The literature on Internet governance and standard-setting has recognized this gap (Brown and Marsden, 2009; DeNardis, 2014, 2015; Cath and Floridi, 2016). These authors have argued that the solution to this is to link human rights legislation to international agreements governing the Internet (Brown and Marsden, 2009; DeNardis, 2014, 2015; Cath and Floridi, 2016). This book unpacks these proposals in the light of current developments in Internet governance.

Brown and Marsden argue for 'prosumer law' and a 'holistic regulation of the Internet' with a number of proposals. They contend that there should be link-up of policy goals between competition authorities which are mainly responsible for regulating the Internet and sector-specific regulators (such as the Federal Trade Commission (FTC) and the UK Office of Communication (Ofcom)). The proposal is that interoperability be enforced through competition law building on case law such as the European Commission's efforts to enforce browser choice.[2,3]

On consumer privacy, Brown and Marsden argue that the main market players should be mandated to agree 'interoperability to permit exit' (2013:185). On social networks, their proposal is that 'competition authorities should impose ex ante interoperability requirements upon dominant social utilities . . . to minimise network barriers' (2013:190). They further propose that there should be regulatory action for transparency, particular of application programming interfaces (APIs), namely software which promotes interoperability, so that

[2] The European Commission fined Microsoft €561 million Euros in 2013 for not complying with commitments to offer users browser choice https://europa.eu/rapid/press-release_IP-13-196_en.htm based upon a 2009 antitrust decision https://europa.eu/rapid/press-release_IP-09-1941_en.htm.

[3] In 2009, via a Statement of Objections, the European Commission put pressure on Microsoft not to bundle Internet Explorer with its personal computer (PC) operating system as it denoted an abuse of a dominant position. The Statement of Objections was based on a 2007 Court of First Instance decision (case T-201/04), upholding a 2004 European Commission competition decision (see IP/04/382) determining that Microsoft had abused its dominant position in the PC operating system market by bundling Windows Media Player with its Windows PC operating system: http://europa.eu/rapid/press-release_MEMO-09-15_en.htm?locale=en.

consumers are aware of its existence as well as its use. Transparency could be attained via procurement power (Brown and Marsden, 2013:198) as well as greater dialogue between civil society stakeholders and standards experts (2013:200).

Brown and Marsden (2013) and Cath and Floridi (2016) argue that 'prosumer' and human rights approaches can be addressed within existing international organizations. As Marsden points out, the 2011 OECD guidelines for multinational enterprises state 'enterprises are encouraged to: support, as appropriate to their circumstances cooperative efforts in the appropriate fora to promote Internet Freedom through the respect of freedom of expression, assembly and association online' (Brown and Marsden, 2013:195). In 2011, the UN Human Rights Council produced Guiding Principles on Business and Human Rights for implementing the UN 'Protect, Respect and Remedy' Framework,[4] with point 12 establishing an annual forum for business and human rights including communications. With this legal foundation in place, Marsden and Brown propose an extension of the WTO's Information Technology Agreement to include the ten principles of the IGF's multistakeholder Internet Rights and Principles Coalition (IRPC).

The IRPC, which is defined as an 'open network of individuals and organisations' based at the UN IGF, is committed to 'making human rights and principles work for the online environment'. It proposed ten Internet Rights and Principles in 2011. The most salient for standard-setting are accessibility (principle 3), privacy and data protection (principle 5), network equality (principle 8), and standards and regulation (principle 9). Principle 9 states that 'the Internet's architecture, communication systems, and document and data formats shall be based on open standards that ensure complete interoperability, inclusion and equal opportunity for all'.

This book demonstrates how there is growing recognition that SDOs must fill the gap in participatory governance, for example with the establishment of the IETF's Human Rights Protocol Considerations Research Group (HRPC) in 2015. Overall, there is a growing recognition that there needs to be consideration of the public interest in protocol development. At the same time, this recognition is pitched against the background of increasing spill-over of policy areas into the standard-setting sphere which are highly politically contested at national levels. Ackrill, Kay, and Zahariadis state that spillover may affect the 'nature of the opening of a (policy) window (2013:876). However, it does not determine its direction. Kingdon saw spillover as setting a precedence for the

[4] A/HRC/RES/17/4.

continued direction of a given policy area (2003:190). This current book instead witnesses changes in policy directions as a result of spillover of existing policy domains to SDO fora. Spillover is instrumental in redistributions of power. It has increased rather than decreased the balance of private sector governance vis-à-vis citizen representation in Internet governance.

Applying the Multiple Streams Approach

In order to investigate SDO decision-making processes, the book applies the MS approach (Zahariadis 2007; Ackrill, Kay, and Zahariadis, 2013, Zahariadis, 2016). Kingdon defined 'agenda-setting' as 'the list of subjects or problems to which government officials, and people outside of government closely associated with those officials, are paying some serious attention at any given time' (1984:3). Although MS was initially presented as an agenda-setting model, Zahariadis expanded it to the decision-making part of the policy process (Zahariadis, 1992, 2003:10; Ackrill, Kay, and Zahariadis, 2013:874). Importantly, he demonstrates that due to the absence of a central authority, policy entrepreneurs become more empowered as they are able to utilize their skills more effectively. As Stone and Ladi note, 'at the global level, the lack of a centralized authority is even more pronounced' (2015:848). The decision-making lens allows us to identify mechanisms which lead to policy decisions. This is particularly enhanced 'in the presence of loose couplings with few rules and generous ambiguity' (Zahariadis, 2008:530). Ambiguity in this case is an organization property.

The framework categorizes how policy problems get on to the agenda, how and where constellations of actors carry out their search for solutions, and how the selection of options occurs. It allows us to identify the framing and coalition-building strategies of important actors, their ability to influence the agenda, and the mechanisms that lead to eventual decisions. The case study chapters explore how and why policy solutions appear and proceed through the agenda of international fora, while other solutions are discarded. Kingdon's focus on events, changes of indicators, and feedback will be utilized, as well as time limitations on decision-making (Kingdon, 2011:185).

Adapting the MS approach, the SDOs under study will be viewed as arenas under which three 'streams' of activity occur: problems, policies, and politics. According to MS, policies are made when three streams—problems, policies (solutions), and politics—are coupling linked by policy entrepreneurs during open windows of opportunity. The 'problem stream' catalogues regulatory problems under discussion. The 'policy stream' identifies solutions proposed by different actors. The 'politics stream' explores the overlapping pressure exerted on actors in self-regulatory fora by states and rival organizational contexts. This takes into consideration the national mood, interest groups, and state preferences (Kingdon, 2011).

Coupling occurs only when assumptions about ambiguity are verified, otherwise the MS framework cannot apply. These assumptions are: variation over time of the fora where the search for alternatives is carried out; and there is fluid participation of participants and actors across fora. In this latter case, actors start participating in one forum then the decision moves to another forum with different actors. This can be observed in the case of protocol development of the Internet of Things (IoT) in the IEEE and Encrypted Media Extensions (EME) in the W3C where spillover has widened participation; cases of unclear technology where participants do not understand how their organization transforms inputs into outputs; and in instances of bounded rationality.

The book investigates the MS framework's main elements within the three streams, namely policy windows, coupling, and policy entrepreneurship. It enables us to see whether and how actors behave as policy entrepreneurs within SDOs and when and where policy windows open and coupling occurs. It assists a wider audience beyond those involved in SDO technical decision-making in learning how attention is manipulated, entrepreneurship used, and options are selected. The MS framework is used to assess decision-making within key Internet governance SDO fora. This enables us to identify policy entrepreneurs and, in particular, dominant actors in the process, particularly from the private sector. With SDO decisions and our case studies, the MS framework works as a lens through which to assess evolving policies within SDO fora. Given the structure of SDOs, the decision-making phase can be integrated into the agenda phase.

Public policy literature is highly relevant for understanding SDO decision-making. Building on incrementalism, public policy scholars theorized that decision-makers worked according to standard operating procedures (DiMaggio and Powell, 1983; March and Olsen, 1984). Charles Lindblom's initial work on incrementalism is based on the premise that regulators will build from year to year on pre-existing policies and that any change is incremental. Most work on incrementalism was based on the evolution of government budgets wherein each annual budget differed only marginally from the previous year.

The book identifies this pattern in SDOs where there has been little change in their modus operandi since they were first established. Institutionalists maintain that organizations are able to continue to operate without change even in the face of transformations in the external environment or pressure from external actors. As outlined earlier, abstract ideals such as human rights and the public interest have been raised within SDO fora, but have not yet managed to maintain a foothold within the decision-making process. In these settings, pressure for change is normally weathered because public attention is normally short-lived and, in the case of technical decision-making, seldom comes to the fore.

This literature contends that participants in organizations choose established rules and procedures over change. Jones argues that the preference for stable rules is enhanced when the external environment is complex and uncertain and when there is overlapping competence for decision-making between different institutions (2001:142). Baumgartner and Jones showed that, in a small number of cases, seismic change can occur; however, policy-making on the whole is gradual (2009:289). Incrementalism has been heavily criticized. Authors have argued that if incrementalism were the norm, this would result in an even growth in policy-making across regulatory areas. However, clearly this is not the case.

Kingdon's key contribution in the MS approach was to return to the source of ideas for policy solutions. Kingdon challenged the phrase 'an idea whose time has come' (1984). He argued that ideas do not simply appear on a given agenda but arise due to a long process of negotiation. Kingdon questioned why some ideas capture the attention of decision-makers and others do not. He distinguishes between an agenda and alternatives to that agenda (Kingdon, 1983:4).

Ideas form solutions to policy problems within a given policy network. Kingdon theorized that policy problems could be 'framed' in many different ways in cases of ambiguity. However, in the 'competition for attention', problems and their solutions rarely come top of a policy agenda. Decision-makers work with imperfect information and a limited time frame. Choices are not altogether 'comprehensively rational' or linear. Solutions are seen as independent from the identification of a given problem and the choices are divided into three separate streams.

Kingdon's research was inductive. In investigating how given problems entered an agenda, Kingdon conducted large-scale interviews with decision-makers in the fields of health and transport policy. The agenda-setting theory developed from the interviews carried out. Kingdon was able to discover how issues make their way onto an agenda through 'visible' participants (Kingdon, 1995:197–9). These participants are normally politicians, members of a committee, or members of a prominent elite. Their attention is rationed.

In the second edition of his book, Kingdon expands on the MS framework to distinguish between the agenda-setting phase and the alternative specification phase of policy-making (Kingdon 1995:4). The latter deals with framing and formulation of policy issues by 'invisible' players. These actors consist of academics, bureaucrats, experts, legislative staff, lobbyists, and those writing for specialist journals with invested interests in the policy process, even if the agenda is being formulated elsewhere. At this level, policy is defined according to a list of policy alternatives. Hence, decision-making is formulated via two different processes entailing different actors and alternatives.

The separation of agenda-setting and alternative specification has an overall effect on decision-making. The different layers of actors do not work together to identify problems and offer solutions. Incrementalism is therefore disrupted as rather than working on stepwise adaptions to existing policies, the agenda is influenced by external agendas. In this respect, Kingdon was drawing on the 'garbage can' model wherein decision-making is not a rational, linear process (Cohen, March, and Olsen, 1972). 'Organised anarchies', the organizations where decision-making takes place are seen as garbage cans into which random combinations of problems, solutions, participants, and choice opportunities result in decision-making (1972:3). The decision is not the product of a formal process or in the interest of any given participant but rather a result of timing and chance. It is instead a melding of 'several relatively independent streams within an organization' rather than just one (Cohen, et al., 1972:3). Policy problems are ambiguous and decision-makers struggle to understand and produce workable solutions under time constraints. Solutions are often a product of timing and the urgency of policy-makers to demonstrate action.

Kingdon built on this model by delineating the decision-making process to three streams with their own separate actors and modes of operation. He then theorized the 'window of opportunity' by way of which different streams can come together and issues are adopted on to an agenda.

The problem stream provides the background to a policy process where problems surface which require attention. There are, however, no clear indicators under which problems can be prioritized. Attention is 'framed' or defined by actors who wish to bring a policy problem to the fore. Unforeseen developments act as 'focussing events' which bring public attention to a problem (Kingdon, 1984:94–5). Kingdon cites transportation accidents and startling government statistics such as a sharp rise in bankruptcy rates and legislative deadlines. Less evident focusing events may take the form of 'a powerful symbol' or 'the personal experience of a policy maker'. Kingdon uses the example of the 'CAT scanner problem' as an example of the high cost of healthcare or the Washington Metro as an example of the high cost of transport (Kingdon, 2013:97). Ackrill and colleagues point to bank implosions in Ireland and political conflict in Georgia (Ackrill, Kay, and Zahariadis, 2013). These events warrant immediate attention from decision-makers and bring pre-existing solutions into policy streams.

Actors in the problem stream tend to be those who are not permanent members of a policy community. Problems can also dissolve with legislation or disappear entirely from the stream. Kingdon distinguishes between definitions of problems and conditions of problems. The latter, such as bad weather or untreatable diseases, cannot be acted upon (Kingdon, 2013:109). However, conditions can become manifest in problems if solutions are identified. This process is a political one, however, which is affected by subjectivity.

The policy stream is one where experts engage in ongoing discussions about different policy options. Actors such as civil servants, legislative staff, and lobbyists operate invisibly under 'alternative specification', teasing out different policy alternatives based upon their areas of expertise. Kingdon defines this as a '"policy primeval soup" in which specialists try out their ideas in a variety of ways' (2013:19). In this stream, impractical solutions are filtered out. Each policy community competes for the limelight, rehearsing arguments, producing policy documents, and building networks. Ideas are revised and modified by a series of actors within issue networks. It is here that 'policy entrepreneurs' engage in advocacy. They are continually looking for opportunities to promote their own policy solutions for potential policy problems by exploiting key events. According to Kingdon, '[e]ntrepreneurs advocate their proposals, as in the softening up process in the policy stream' (2013:183). In this respect, there is a tendency of 'solutions to chase problems' rather than the other way around, as 'advocacy of solutions often precedes the highlighting of problems to which they are attached' (Kingdon, 1984:215). This goes some way to explaining why some solutions do not fit well with the problems to which they are assigned.

The politics stream is one where 'visible' policy actors such as politicians, their staff, legislative committee members, and perhaps the media define the agenda. It gives public presence to a policy problem. Agendas are broad and hampered by time and budget constraints. Policy problems are therefore influenced by political careers, party strategies, and public discussion. Politicians are not specialists and are influenced by party agendas, public reputation, and the 'national mood'. Indeed, in his study of foreign policy, Zahariadis divides the politics stream into 'three elements: the national mood, pressure group campaigns, and administrative or legislative turnover' (2005:23). These elements act as a gauge which governs the way in which politicians pay attention to a particular issue. In the politics stream, interest groups can prompt the inclusion of a policy solution onto an agenda. Public dissent can remove an issue entirely (Zahariadis, 2005:24). When there are conflicting views, politicians seek balance between different interests (Zahariadis, 2005:108). For Zahariadis, the politics stream is the most salient of all the streams but it is not necessarily an indicator of policy change. The way in which attention is rationed is dominated by policy entrepreneurs. The politics stream is problematic for use in this book given that there is no role for politicians in SDO fora. For this reason, we adapt the politics stream to decision-makers within the fora. 'Visible' actors as seen in the book are senior SDO staff such as Tim Berners-Lee within the W3C, area directors (ADs) within the IETF, and members of deciding committees within the IEEE. The book's chapters assess whether the public mood is reflected in SDOs in cases of data sharing, copyright, and privacy.

The MS literature stipulates that policy entrepreneurs are key to creating policy windows. At the last stage of the process, policy windows appear when the three streams are coupled or join together at key junctures in time. Kingdon referred to these as 'policy windows' as 'this coupling is most likely when policy windows—opportunities for pushing pet proposals or conceptions of problems—are open' (2013:20). In this scenario, the political sphere (politicians, political decision-makers) decide that there is a problem to solve and the policy stream has produced viable informed solutions. Policy windows normally occur for a short period of time depending on the issue at hand. At this point, entrepreneurs link their solution to the current problem. Entrepreneurs coordinate 'negotiation among people and making the critical couplings' (Kingdon, 2013:183). Success is often dependent upon access to decision-makers. It is during open policy windows that policy entrepreneurs link the three streams together (problems, policies, and politics) (Kingdon, 2013:873). Policy entrepreneurs act as 'power brokers and manipulators of problematic preferences and unclear technology' (Zahariadis 2003:155).

Ackrill and colleagues point to three factors which determine which solutions get attention (Ackrill, Kay, and Zahariadis, 2013:872). First, time is a valuable resource in decision-making as decision-makers cannot address all problems and cannot accept outcomes which are not always optimal but which do satisfy the problem. Second, 'means and ends, solutions and problems are generated independently of each other' (Ackrill, Kay, and Zahariadis, 2013:872). This means that framing determines how solutions meet problems within the MS process. Third, ambiguity means that entrepreneurs can manipulate decision-making.

A key criticism of the MS approach includes its inadequacy for quantitative research which takes a top-down rather than bottom-up approach. The independence of the streams has also been called into question. However, it is still in use by many academics. Its strength is its flexibility as an explanatory framework. A number of theories such as punctuated equilibrium have built on the MS framework, which centres on the allocation of attention and disproportionate information within the policy process.

Indeed, the original MS framework contains underlying assumptions about the policy process. Importantly it was originally developed to explain national-level policy-making overseen by government processes. Scholars such as Bache have applied it to the European Union to demonstrate how Member States operate within an MS framework (Bache, 2012). Peters noted that application of agenda-setting theories to the European level is significantly different than the national level due to the 'existence of a number of points of access, of a larg(er) number of influential policy advocates, and a wide range of policy options' (1994:11). The European Union fits this description due to its high institutional complexity. This is less true of SDO fora because access is severely restricted and restrained by cost, time, and expertise required of participants. So, rather than opening the process to more actors, it closes off access to the technically apt. However, Zahariadis argued that the MS framework explains decision-making well when issues are highly complex but institutional complexity is low (2013:814). Accordingly, this book departs from previous (national and EU) work by applying the MS framework to the international level and exploring its usefulness in cases of governance but not government.

Identifying Mechanisms

In order to increase the impact of policy analysis on 'real-world' decisions, it is argued in the book that it is fundamental to expose the mechanisms at work, not just the 'streams' or the structural conditions. Thus, we explore mechanisms in assessing how agendas evolve into selection of policy solutions, given certain problems and political contexts.

For this purpose, this book enriches the MS approach with the literature on policy mechanisms (Gerring, 2006; Falleti and Lynch, 2009) and international self-regulatory fora (Drezner, 2004; Goldstein and Steinberg, 2010, Mattli and Büthe, 2013). It observes which mechanisms act as focusing events that trigger policy solutions within each stream. Mechanisms are also important in the identification of coupling of streams and changes in issue framing. Stream coupling is crucial to the opening of policy windows through which alternative policy solutions can travel. Coupling is a major component of the MS framework where entrepreneurs package solutions which are then 'sold' to decision-makers (Ackrill, Kay, and Zahariadis, 2013:873).

The book identifies key mechanisms which trigger the adoption of policy solutions during decision-making, drawing on Kingdon's approach and the public policy literature (Benkler, 2006; Farrell, 2006; Mattli and Büthe, 2013). Four main mechanisms (exogenous, institutional, collective, and individual) are identified which are outlined below, distinguishing between macro (exogenous and institutional) and micro mechanisms (collective and individual) (Gerring, 2010). The book develops ideas from existing literature by observing how mechanisms work within different MS streams. The premise is that mechanisms will behave differently within different streams which will contribute to the application of the MS approach. This is analysed on a comparative basis in Chapter 12.

The first macro mechanism is exogenous. Exogenous mechanisms generated by systemic shocks and technological change can strengthen the ability of actors to alter SDO agendas. Chapter 4 in particular assesses how the 2013 Snowden revelations acted as an external shock which provided a window of opportunity to couple a solution (the Quick UDP Internet Connection (QUIC) standard) within the problem stream (Transport Layer Security (TLS) flaws). Overall, it shows how Snowden strengthened the ability of certain actors to alter the IETF agenda vis-à-vis historical actors. The QUIC approach adopted was supported by civil society groups and the public interest was promoted via a hybrid solution to privacy and data protection. Chapter 11 views the IoT as a systemic shock opening a window of opportunity for SDO reform in intellectual property rights. The increase in the need for interoperability by the IoT, which enables the linking of devices across networks never before envisioned, led to a push for SDO revisions to their intellectual property rights (IPR) policies. This was flanked by pressure from the United States and the European Union.

The second macro mechanism is institutional, relating to external pressures. The institutional mechanism of external pressure can lead to agendas favouring corporate-led proposals. As stated earlier, authors have demonstrated how states utilize corporations as 'points of control' to attain public policy goals in Internet governance (Benkler, 2006; Farrell, 2006). Chapter 9 discusses the relevance of 'points of control' when analysing state surveillance. State control is identified as the restriction in cryptography research, the interception of TLS traffic by security agencies, the development of protocols by intelligence agencies, and the use of 'back door' technologies. The chapter shows that these tactics have constrained SDO decision-making. Chapter 11 shows that state 'points of control' played their part in this as well. The IoT provided a window of opportunity for change but it was pressure from the European Union and the United States in particular which drove SDOs to change their IPR policies.

Chapter 8 explains how different regulatory pressures in the United States and the European Union have acted as exogenous mechanisms in the case of radio spectrum. State preferences are at the heart of the politics of white-space IoT developments and account for policy solutions developed for dynamic spectrum access (DSA) standardization within the IEEE and the Third Generation Partnership Program (3GPP). In this respect, authors observing exogenous mechanisms demonstrated how the size and importance of a state's market are key determinants of corporate actor behaviour within international fora (Drezner, 2004; Goldstein and Steinberg, 2010).

The first micro mechanism is collective in the creation of constituencies and boundaries, gradual change, layering, and policy drift (Falleti and Lynch, 2009:1150). Policy drift refers to situations in collective decision-making where stasis in decision-making exists internally to an organization when the external context changes (Hacker, 2005). In this situation, a given organization suffers from lack of reform or adjusts poorly to new environments. It often results in what Hacker referred to as 'layering' where new rules are introduced on top of or in addition to old rules. According to the historic institutional literature, layering occurs where there is a low level of regulatory oversight or implementation is only voluntary. New schemes are formed even though the older ones continue to exist. This leads to the creation of new internal constituencies and boundaries (Falleti and Lynch, 2009:8). Chapters 5 and 6 examine this process. Chapter 5 explains how layering occurs with the development of the Hypertext Markup Language (HTML) standard within the Web Hypertext Application Technology Working Group (WHATWG) and the W3C. This led to a division of labour and the opening of new spaces for the spillover of work to the W3C from the national sphere. In the development of Encrypted Media Extensions (EME), action is limited and shaped by external legal environments. Chapter 6 focuses on policy drift with the development of 802.11ax within the IEEE. Both chapters conclude that these mechanisms have led to eventual actor-led ideational change within the main SDOs, the W3C and IEEE.

The second micro mechanism is individual governing actor behaviour and learning, perceptions, the social construction of roles, and how opportunities emerge from dense interaction and social control (Falleti and Lynch, 2009). As Falleti and Lynch observe (2009:18), actor-led ideational change can be observed when decision-making is divided into time segments where the relevant context is constant. This can lead to coupling of policy solutions within the problem stream as the social construction of roles governing actor behaviour and learning acts as a mechanism for policy change.

Chapter 7 describes incidents of actor-led ideational change. It assesses the Do Not Track standard (DNT) as a response to tracking behaviour that has sought to ensure that users can choose whether or not they are tracked online. It focuses on the extent to which different interests and stakeholders converged on its conception, and development with a specific interest in the role of civil society in the W3C. This, of course, occurred in the broader (self-) regulatory, legal, and political environment within which the DNT standard has been developed, agreed, and implemented. To this end, the chapter pays attention to the appearance of the problem, how a consensual agreement (Candidate Recommendation) was arrived at in the W3C, missed windows of opportunity, and the extent to which the DNT standard has been adopted and enforced. The chapter focuses particularly on how the evolving policy and politics streams in the European Union and the United States relating to privacy have exerted pressure on the relevant actors (in particular advertising networks) to honour DNT.

Chapter 10 continues with the focus on actor-led ideational change within the W3C and IETF promoted by public campaigns by the digital rights group the Electronic Frontier Foundation which highlighted the evasion of personal privacy. It demonstrates how public interest initiatives can be inserted into protocol development with a case study on browser fingerprinting. In this respect, actor-led ideational change can bring about a change in SDO participant behaviour based on underlying societal norms concerning the right to transparency and privacy.

Methodologically, the book assesses the behaviour and influence of civil society organizations in order to ascertain how the public interest is promoted across the cases chosen. Each chapter will:

1. identify how and why certain policy solutions appear and proceed through the policy agenda of the relevant international self-regulatory fora under investigation.
2. ascertain the degree of policy entrepreneurship and agency in pushing the adoption of proposed solutions within selected case studies and the SDO under discussion.
3. establish the mechanisms which enable or constrain CSOs in promoting policy solutions.

4. make a contribution to the provision of a new conceptual model of actor participation in Internet governance.
5. determine and analyse the public policy implications of civil society participation in Internet self-regulatory fora.

To this end, the book addresses the following key research questions: What are the key political problems under discussion within the Internet's self-regulatory fora? This enables the analysis of the problem stream. What solutions are proposed by identifiable actors within fora? This enables analysis of 'policy streams' with investigation of solutions which are framed within the streams. Is the United States or the European Union or other states steering actors within 'politics streams'? Which strategies enable actors to couple the three streams? Why and under what conditions are they most effective? Which mechanisms enable or constrain them? What are the public policy outcomes?

Is there any civil society representation and does this address public interest concerns?

3

Internal Governance of the IETF, W3C, OASIS, and IEEE

Structure, Decision-making, and Internationalization

This chapter outlines the key features of standard-developing organizations (SDOs) involved in Internet governance. Crucially, these fora are voluntary in nature and have no formal state representation. The chapter outlines their governance structures and most salient areas of work. To this end, it charts the evolution of each SDO, their main purpose, function, and central decision-making processes. The latter can be affected by matters such as the number of actors involved in the process and the degree of difference in viewpoints expressed, the complex and interdependent nature of the technologies under examination, the range of technical alternatives put forward, and the technical and economic significance of the standard in question (Simcoe, 2007:262). The chapter highlights the differences as well as the degree of synergy and collaboration between each of the SDOs.

The chapter's central purpose is to provide a governance context through which to understand the decision-making process in each of the standards investigated in the case study chapters. It also explains how procedures differ between the fora and how moving goalposts and high barriers to entry make it difficult for civil society to participate. The chapter identifies differences in EU versus US approaches to self-governance and key drivers of current policy development such as internationalization and the Internet of Things (IoT). It explores how problems of security, capacity, the pace of technological development, internationalization, and intellectual property rights (IPR) cut across all ongoing work. The focus is on the principal SDOs engaged in the development of the Internet's architecture, namely the Internet Engineering Task Force (IETF), the World Wide Web Consortium (W3C), the Organization for the Advancement of Structured Information Standards (OASIS) and the Institute of Electrical and Electronics Engineers (IEEE) (see Table 3.1).

In the 1980s, the computer networking market was dominated by governmental (the Advanced Research Projects Agency Network (ARPANET)

Global Standard-Setting in Internet Governance. Alison Harcourt, George Christou and Seamus Simpson, Oxford University Press (2020).
DOI: 10.1093/law-oso/9780198841524.001.0001

Table 3.1 Organizations involved in cross-national Internet governance

Registries	Intergovernmental fora	EU and national organizations	Key standard-developing organizations (SDOs)
APNIC, ARIN, CENTR, ICANN, LACNIC, RIPE-NCC, national registries (Nominet, Verisign, etc.)	CoE, EU, ePol-NET, ISO, ITU, OECD, UN, WIPO, WTO	CEN, CEPT, ECTRA, ETNO, ENISA, ETSI, NATIA	IAB, IEEE, IETF, Internet Society, IRTF, OASIS, 3GPP, WHATWG, WC3

Source: Compiled by authors.

development within the US Defense Advanced Research Projects Agency (DARPA); CYCLADES at the French Institut de Recherche en Informatique et en Automatique (IRIA)), academia (Computer Science Network (CSNET)), or large firms such as IBM with its System Network Architecture (SNA) and DECnet of Digital Equipment Corporation (Bing, 2009:33; Alverstand and Lie, 2009:127). Lawrence Roberts created ARPANET by moving away from circuit switching to package switching technology in 1969, linking up computers at four universities: the University of California, Los Angeles (UCLA), Stanford, University of California, Santa Barbara, and the University of Utah (Lee and Bing, 2009; Brown and Marsden, 2009; Brown, 2013:7; DeNardis, 2014; Contrares, 2016:867; Musiani and DeNardis, 2016:33). Robert Kahn from DARPA, working together with Vint Cert from Stanford University, built on this technology to create the Transmission Control Protocol (TCP) and Internet Protocols (IP). TCP/IP became the main standard for the Internet.[1]

TCP/IP use was facilitated in particular by the Sun Microsystems UNIX operating system in the early 1980s, and its use in linking local area networks (LANs) to ARANET for the communication of personal computers (PCs). IBM dominated the market in the early 1980s with launch of the PC. These used Intel chips and were linked up by Ethernet developed by Xerox. Local area networks were pioneered by Cisco.

An attempt to provide an international standards framework for interoperability was launched by the International Organization for Standardization (ISO) and the Telecommunication Standardization Sector (ITU-T) of the

[1] TCP/IP was standardized formally in RFC 791 and RFC 793 in 1981.

International Telecommunication Union (ITU) in 1977. Most specifications at that time were proprietary in nature (Brown, 2013:17; Besen and Sadowsky, 2016:212). The main alternatives to TCP/IP were the Open Systems Interconnection (OSI), DECnet, and SNA/APPN protocols. The Internet Architecture Board's (IAB) initial decision to adopt the Open Source Initiative's (OSI) ConnectionLessNetwork Protocol in place of IP at the IETF in 1992 met with opposition within the IETF. As detailed by Yates and Murphy, Vint Cerf 'became a hero by stripping off his jacket, vest and shirt during his talk to reveal a T-shirt reading "IP on Everything"' and David Clark gave his time honoured talk 'We reject: kings, presidents and voting. We believe in: rough consensus and running code'" (Yates and Murphy, 2019:250). The take up of TCP/IP, which was open source software and free to use, was so widespread that it became the standard networking protocol for the Internet (Bing, 2009:27; Besen, and Sadowsky, 2016:216). This was important for establishing the principle of interoperability and the delineation between the different protocol layers.

The Internet is made up of layers. IP defines how data should be packaged and addressed and how data should travel through the Internet stack. Lessig identified three layers: the code layer, the content layer, and the physical layer (2001). Benkler refers to three layers: physical infrastructure layer, logical layer, and the content layer (2006). Bing and Lee referred to six layers for the Internet (content, application, transport, internet, link, and physical). This book refers to four layers, using the model offered by Contreras (2016).

Layer 1 comprises the network, which refers to physical transmission and datalink technologies. This consists of wired (i.e. Ethernet, Services Description Language (SDL), and Integrated Services Digital Network (ISDN)) and wireless (2G/3G/4G/5G) connection standards. Layer 2 constitutes Internet protocols (IPv4/6, etc.) which enable the identification and location of computers and devices on networks and which routes traffic across the Internet. Layer 3 incorporates Internet protocols (Transport Layer Security (TLS), Quick User Datagram Protocol Internet Connection (QUIC)) which govern handshaking and the transmission of IP packets. In Layer 3, TCP and UDP come into play which provide host-to-host connectivity. The top layer 4 handles HTML descriptor language and software interfaces.

Each layer has corresponding SDOs which are associated with them that ensure interoperability. Interoperability is crucial for the Internet to function properly. Standards within SDOs are agreed via broad consensus and often technical merit. However, consensus, transparency, and procedures differ from one forum to the next. There is a wide range of SDO fora involved in Internet

Table 3.2 Internet standardization stack

Layer	Standards	SDOs
4. Application	XML, HTTP, HTML, IMAP, POP, MIME, SMTP, Telnet, TLS/SSL	IETF, OASIS, IETF, W3C
3. Transport	TCP, UDP, TLS, QUIC	IETF
2. Internet	IPv4, IPv6, ICMP, ECM, Seamoby	IETF
1. Network	Ethernet, IEEE 802. 11, ISDN, DSL, WiFi, 3G/4G/5G	IEEE ETSI 3GPP

Source: (Adapted from) Contreras, 2016:856.

governance including those at national (e.g. UK National Standards Body), regional (e.g. European bodies such as CEN, CENELEC, and the European Telecommunications Standards Institute (ETSI)) and international (e.g. the International Organization for Standardization (ISO), the International Electrotechnical Commission (IEC), IEEE, IETF, and W3C) levels to smaller specialized consortia such as the 3rd Generation Partnership Program (3GPP). The IETF, the W3C, and the IEEE are the main self-regulatory fora operating at the international level underpinning interoperability for the Internet (Table 3.2).

The Internet Engineering Task Force

The IETF is the most active organization in Internet standard development (Request for Comments (RFC) 4677) and arguably the most important in the first three layers. It standardized TLS, SSL, QUIC, IPv4, and IPv6 protocols. A key goal of the IETF is to provide an organizational context and support for Internet standards-making. Established in 1986 from humble beginnings, when its first meeting was attended by only twenty-one people (Simcoe, 2007:260; Alvestrand and Lie, 2009:127; DeNardis, 2015:74), the IETF and IAB developed out of the initial collaborative work of DARPA and universities. In its early days, the IETF subsequently catalogued most Internet standards produced before its inception (Gencher, 2012). However, its roots are in the 1970s and work of engineers in ARPANET where Vint Cerf and David Clark founded the Internet Configuration Control Board, which later became the Internet Activities Board and thereafter the IAB (DeNardis, 2014).

Reflecting its academic roots, the IETF's focus was built around technical merit, transparency, and broad consensus. Simcoe argues that the lack of commercial presence in the earlier years of the IETF's history was to some degree due to the US National Science Foundation's acceptable use policy (ended in 1991), which excluded commercial use of the network (Simcoe, 2009). This continued even though members from the commercial sector quickly outnumbered academics and government representatives from the mid-1980s onwards (Simcoe, 2007:264). According to the IETF's mission statement, work is based upon 'engineering quality'.[2] Participants are viewed as individual representatives, not representatives of corporations or organizations, and technical progress is valued over the organizational priorities. There are no fees for membership. The Internet Society (ISOC) was created some years later in 1992 as the administrative arm of the IETF (see following).

The IETF's current mission was last updated in 2015 (RFC 3935) and states:

> the mission of the IETF is to produce high quality, relevant technical and engineering documents that influence the way people design, use, and manage the Internet in such a way as to make the Internet work better. These documents include protocol standards, best current practices, and informational documents of various kinds.

A core concern is to ensure that standards produced are tested in advance to minimise risk to the efficient functioning of the Internet. For this reason, the IETF encourages 'Hackathons' at IETF meetings for developers to collaborate, share code, and practise different implementations of IETF standards. Otherwise, participation is mainly organized online and involves open exchange within working groups. Most documents and all standards and protocols are available on the IETF website. Anyone is welcome to join a working group. In this way, the IETF is more open than the W3C and IEEE, which demand fees and operate less-accessible working groups.

Nonetheless, most IETF participants are from large corporations and key proposals are concentrated into a small group of companies which sponsor individuals' participation in the form of time and travel to annual meetings. There are experts from around 130 companies represented at the IETF but the leading participants are from Huawei, Ericsson, Juniper, Cisco, and Google. Cisco, Huawei, and Cloudfare work on content delivery networks (CDNs) signalling a high representation from US companies (see Figure 3.1). Cisco

[2] https://www.ietf.org/about/mission/.

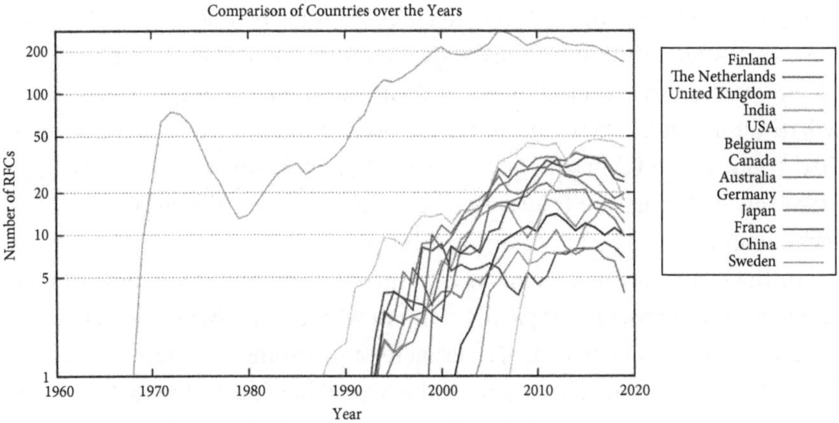

Figure 3.1 IETF RFC production: comparison of companies over the years.
Source: Jari Arkko http://www.arkko.com/tools/rfcstats/companydistrhist.html

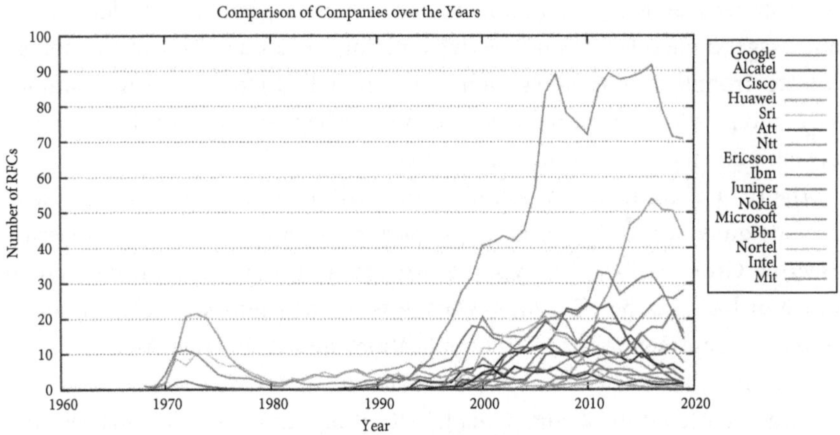

Figure 3.2 IETF RFC production: comparison of countries over the years.
Source: Jari Arkko https://www.arkko.com/tools/rfcstats/countrydistrhist.html

specializes in LAN technology and Voice over Internet Protocol (VoIP). It publishes the most RFCs annually (about seventy to ninety per year) with Huawei coming second (around fifty 50 proposals per year) (see Figure 3.2).

Company presence within SDOs does not necessarily mean that they are the most influential within the Internet stack. Standards are only valid if implemented. Some companies send many representatives to IETF meetings annually and produce a high number of technical standards. Other companies rarely send anyone but this doesn't correlate necessarily to influence. Facebook sends three or four people per year to the IETF. Amazon does not send anyone, however it affects the process via the standards it adopts. These groups arguably support standards produced by other companies. Intel, Google, Amazon, and Microsoft are most dominant in the licensed spectrum domain (Greenstein, 2017). Google and Amazon are active in Wifi. with Amazon, in many interviews, alleged to not be transparent about its Amazon Web Services (AWS) technology.

An IETF standard is created when a specification undergoes development and review. The process is typically complex where a number of factors such as ensuring high technical quality, catering for a wide range of interests, securing broad consensus, and calculations of the overall benefits to the Internet community are important (IETF RC 2026, section 1.2). The IETF does not have a board of directors nor any official members. Rather, it describes itself as 'a loosely self-organized group of people who contribute to the engineering and evolution of Internet technologies' (IETF 2012:2). In so doing, it specifies the development or usage of protocols; recommends protocol standardization and usage to the Internet Engineering Steering Group; assists with technology transfer from the Internet Research Task Force (IRTF) to the Internet community; and provides a forum for information exchange between a diverse range of commercial players, users, and researchers.

The IETF is composed of technical areas led by Area Directors (ADs) who provide guidance to the working groups and sit on the Internet Engineering Steering Group (IESG).[3] They serve two-year terms and come from a broad range of backgrounds. In 2019, work was divided into six technical areas. ADs were representatives from (Applications and Real-Time Area) Huawei, Isode, and Mozilla; (Internet Area) Kaloom and Cisco (Operations and Management), Equinix, and Google; (Routing Area) AT&T, Huawei, and Nokia; (Security) CERT/SEI[4] and Akamai Technologies; (Transport Area) Ericsson and (Liaison and ex-officio) members from Google, the University of California's Information Sciences Institute, Association Management Solutions, and the Internet Assigned Numbers Authority (IANA).

[3] https://www.ietf.org/about/groups/iesg/members/.

[4] For example, the AD for security works for the Software Engineering Institute, the primary mission of which 'is to support the defense of the United States' https://www.sei.cmu.edu/about/what-we-do/index.cfm. The AD for security leads IETF working groups on Security Dispatch, Automated Certificate Management Environment (ACME), the IoT, Public Notary Transparency, Token Binding, the Web Authorization Protocol, Interface to Network Security Functions, EAP Method Update, Security Automation and Continuous Monitoring, Limited Additional Mechanisms for PKIX and SMIME, Remote ATtestation ProcedureS (RATS), Security Area Open Meeting (SAAG), Security Area Directorate (SECDIR), the Tools Team (TOOLS), and Privacy and Security (PRIVSEC).

The IESG approves proposals to create new working groups and plays the key role in approval of proposed standards and their advancement from 'Proposed Standard', 'Draft Standard', and 'Standard'.[5] Once reviewed by the IESG and published, these standards never change. Instead, they become obsolete if the specification is replaced by a subsequent standard. The most fundamental standards for the Internet are TCP, IP, Border Gateway Protocol (BGP), UDP, Domain Name System (DNS), Simple Mail Transfer Protocol (SMTP), and HTTP (Alverstand and Lie, 2009:133). However, there are dozens of standards agreed per year. The IETF has published around 300 RFCs annually over the last ten years, peaking with 459 RFCs in 2006.[6] Jari Arkko, who was chair of the IETF from 2013 until 2017, ran an IETF statistics tool which tracked the number of RFCs published by the IETF over time.[7] The tool documented the production of RFCs between 2013 and 2018 (see Figure 3.3). RFCs (Request for Comments) are not necessarily finalized standards. A commonly heard phrase at the IETF is 'the Internet runs on Proposed Standards' (Alverstand and Lie, 2009:130).

The IESG takes charge of the technical management of IETF activities through rules and procedures ratified by the ISOC Board of Trustees. It is the central coordinating body of the organization made up of ADs selected by the Nominations Committee (NomCom) and plays an important role in reviewing RFC proposals. The IESG decides whether an IETF Working Group has produced an outcome which has secured the famous 'rough consensus' on which IETF decision-making is based. In 2019 there were twenty IESG Members.

The NomCom is established annually by requesting volunteers from the membership followed by a random selection process.[8] The Chair of NomCom is appointed by the Internet Society President. The NomCom is critical to the IETF in that it selects potential members for the IESG, IAB, and Administrative Oversight Committee (IAOC). In addition to NomCom, the IETF has a Secretariat which, among other things, ensures the accurate maintenance of the organization's Internet Drafts directory and a Trust responsible for licensing the IETF's intellectual property. The IAB is both a committee of the IETF and an ISOC advisory body. With regards to the IETF it has a long-term planning and a new activities monitoring role. It has an additional role of implementing 'IANA considerations' within the IETF.[9] Beyond this, the IAB

[5] https://datatracker.ietf.org/doc/rfc2026/ last accessed 3 May 2019.
[6] https://www.ietf.org/standards/rfcs/.
[7] http://www.arkko.com/tools/rfcstats/pubdistr.html.
[8] https://datatracker.ietf.org/doc/rfc7437/.
[9] https://tools.ietf.org/html/rfc8126.

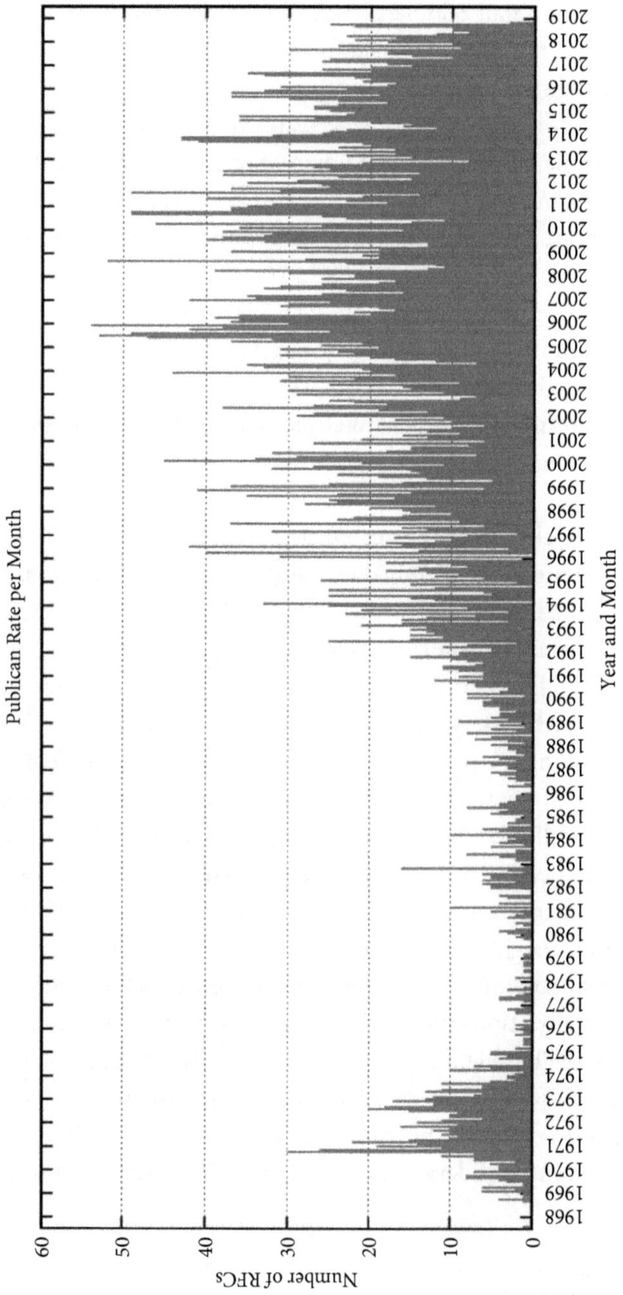

Figure 3.3 RFC publication per year.

Source: Jari Arkko http://www.arkko.com/tools/rfcstats/pubdistr.html

runs the RFC Series Oversight Committee (RSOC) and has oversight of IETF external liaisons. There is also the IETF's Administrative Oversight Committee which is a voluntary body chosen by the IETF NomCom, appropriate ex-officio members of ISOC, and the IETF leadership.[10]

The IETF holds three meetings per year in order to undertake face-to-face progress on ongoing work, and these take place according to its designated charter provisions regarding the organization's scope and objectives. Working groups (WGs) tend to form as a result of so-called Birds of a Feather (BOF) meetings (approved in advance by an AD) where individual experts get together to consider a group charter and work agenda. BOF meetings are held at IETF meetings which are in turn held to garner interest in proposals. BOF meetings are official and listed in meeting programmes.

For example, at the IETF 104 in 2019, the IETF Chair, Alissa Cooper, approved six BOF sessions proposed by the IESG: the Predictable and Available Wireless (PAW); Collaborative Automated Course of Action Operations (CACAO); Broadband Network Gateway Control-Plane And User-Plane Separation (BCAUSE); Key Signing Key Futures (KSKF); Brand Indicators for Message Identification (BIMI); and Stopping Malware and Researching Threats (SMART). BOFs are also organized remotely. Interviews and IETF attendance revealed that there are an increasing number of 'pre-BOFS' organized on an informal basis and side meetings which lead to 'flex-BOFs'. This informal style of governance is not new, however, to the IETF as 'corridor' networking has long been prevalent.

Most of the business of WGs is conducted through mailing lists where consensus on anything agreed at face-to-face meetings must be secured. Securing rough consensus is when

> the chair of a working group determines that a technical issue brought forward by an objector has been truly considered by the working group, and the working group has made an informed decision that the objection has been answered or is not enough of a technical problem to prevent moving forward, the chair can declare that there is rough consensus to go forward, the objection notwithstanding.[11]

The IETF also engages in the peculiar process of humming consensus during meetings, developed to guard against the over-representation of parties with

[10] https://tools.ietf.org/html/rfc4071.
[11] https://datatracker.ietf.org/doc/rfc7282/.

the loudest voices. More broadly, changes in IETF procedures for setting standards are derived through the activities of a WG. The fluid membership nature of the IETF is a key feature of its governance procedures.

After a WG has agreed on a standards document a so-called Document Shepherd takes it forward who is normally, but not necessarily, the WG chair or secretary. An IETF standard (which can sometimes also come from individuals as well as WGs) is published as an RFC. The journey to this point involves a multi-stage process of discussion and review. The Internet Draft document[12] (which has a six-month life) receives comments and is amended subsequently, after which an AD brings the draft to the IESG after reviewing it.

The IESG then creates an organization-wide Last Call. After any concerns expressed by IESG and organization-wide members are addressed to the IESG's satisfaction, the document can be published by the RFC Editor as a proposed standard. Further adjustment to the standard can occur after this point. An interesting feature of the IETF is that most standards in use do not move forward from Proposed Standards to Standards (or 'Internet Standards').

According to the IETF, to become an Internet Standard, a specification must be "stable and well-understood, is technically competent, has multiple, independent, and interoperable implementations with substantial operational experience, enjoys significant public support, and is recognizably useful in some or all parts of the Internet" (RFC 2026).[13] Not all RFCs are standards. There are five types of RFCs: 'standards track', 'best current practice', informational, historic and experimental.[14]

Core issues under discussion at the IETF in 2019 included the web protocol stack (HTTP2, QUIC); security and privacy; real-time communications in browsers; management, orchestration, virtualization, software-and data-model-driven networking; IoT and running code and open source. Interviews showed that civil society groups were most interested in the work on the Atom Publishing Protocol (APP), Domain-based Message Authentication, Reporting & Conformance (DMARC), the DNS-based Authentication of Named Entities (DANE), the Domain Name System (DNS), work related to IEEE WiFi standards, Internet Protocol Performance Metrics (IPPG), Transport Layer Security TLS 1.3, and Extensible Provisioning Protocol (EPP) privacy implementation.

As the Internet has grown and become of prime strategic commercial significance, the workload of the IETF has inevitably become weightier and more

[12] https://www.ietf.org/standards/ids/.
[13] https://datatracker.ietf.org/doc/rfc2026/?include_text=1.
[14] https://www.ietf.org/standards/rfcs/.

complex in character. Much more is at stake. Interviews revealed that there is an internal understanding that the IETF only publishes one specification for a single requirement. This means that, if there are two competing groups, only one specification is adopted. Often competing standards revolve around IPR issues.

The IETF has also shown reluctance to work on protocols not operable for devices not using TCP. An example of this is when it dealt with an extension for the IETF protocol TCP called Q.FlowstateSig.[15] Unable to find AD support within the IETF, the consortium took Q.FlowstateSig to the ITU-T. However, a compromise was sought via the IESG and ITU-T Q5/SG11 liaison committee in that the protocol had a lowest common denominator implementation so only those who wanted to use it had to employ it.

The Internet Society

There is a close relationship between ISOC and the IETF. ISOC houses the IETF, the IAB, the Internet Engineering Steering Group (IESG), and the IRTF. It is registered as a non-for-profit organization in the United States. ISOC was initially created in 1992 to provide financial support for the IETF. Prior to this it had been funded by the National Science Foundation at the Corporation for National Research Initiatives (CNRI), founded by Robert Kahn. There were concerns at the time of the IETF's establishment surrounding its ability, as a non-legal entity, to cope with legal challenges arising from the increasingly strategically important Internet standards-setting process. ISOC was viewed as the organizational answer in terms of both providing an appropriate legal standing and generating financial revenue (Alverstand and Lie, 2009:128; DeNardis, 2014:70). ISOC generates most of its funding via the Public Interest Registry (PIR) which has allocated the .org top-level domain (TLD) since 2002. As Braman notes, Vint Cerf took pains to point out that the 'Internet is for Everyone' in RFC 3271 denoting its public interest role (Braman, 2016:5).

Discussion in IETF and IAB meetings in 1991 led to the establishment of ISOC. Vint Cerf served as the first President of ISOC's Board of Trustees. A key moment in the development of ISOC's core relationship with the IETF occurred in June 1992 when the ISOC Board of Trustees decided to accept the IAB invitation to bring the IAB and its activities into ISOC so that the IAB became a technical advisory group of ISOC. Cerf argued that round this time

[15] https://www.iab.org/documents/minutes/minutes-2011/iab-minutes-2011-08-24/.

'it appeared that long term support for the standards-making activity of the IETF, which had come primarily from research supporting agencies of the US Government (notably ARPA, NSF, NASA, and DOE) might need to be supplemented in the future'.[16] ISOC was proposed as an organization that could draw together and extend the IETF's funding. This meant that the IETF's activities came under this new arrangement. In the second half of 1992, the Process for Organization of Internet Standards (POISED) working group 'reallocated responsibilities for standards decision-making' and made recommendations which became the basis for the working relationships between the IAB, the IESG, ISOC, and the IETF. ISOC's support for the IETF standards process was affirmed at a 1995 ISOC Board of Trustees meeting.

In 2003, an ad hoc advisory committee of the IAB made recommendations on the future administrative needs of the IETF (IETF, RFC 3716). The IETF asked for ISOC's support in a subsequent restructuring which resulted in the establishment of the IASA (see earlier) within ISOC. This led to a streamlining of the IETF's administration and greater individual control of its budget (IETF RFC 4071) which has hovered around US$5 million annually for the past twenty years.[17] Alongside this, the US Corporation for National Research Initiatives and ISOC established a trust for the IETF. This ensured that the IETF (with the IAB and RFC Editor) developed within the ambit of ISOC. Since this time, ISOC has established a voluntary relationship with the ITU.

ISOC was incorporated as a non-profit organization. According to its Charter its aims are:

> To facilitate and support the technical evolution of the Internet as a research and education infrastructure and to stimulate involvement of the academic, scientific, and engineering communities (among others) in the evolution of the Internet. To educate the academic and scientific communities and the public concerning the technology, use, and application of the Internet. To promote scientific and educational applications of Internet technology for the benefit of educational institutions at all grade levels, industry, and the public at large. To provide a forum for exploration of new Internet applications and to foster collaboration among organizations in their operation and use of the Internet.[18]

[16] https://www.internetsociety.org/internet/history-of-the-internet/ietf-internet-society/.

[17] ISOC contributes about half of the IETF annual budget. Other IETF funding is derived from conference registration fees, meeting sponsorships, in-kind sponsorships, event sponsorships, 'Bits-N-Bites', 'Hackathons', and hotel commissions. https://iaoc.ietf.org/budget-and-finance.html.

[18] https://tools.ietf.org/html/rfc2134.

ISOC's board of directors is the Board of Trustees (originally made up of thirteen people) which takes action on a majority vote basis (or unanimity, in the case of an electronic vote), and four of whom are appointed by the IETF. ISOC had four declared priority areas in 2019: building trust; connecting the world; improving technical security; and shaping the Internet's future. In respect of connecting the world, ISOC declared in 2019 that it aims to 'promote policies and build communities to sustain an Internet that is a universally accessible platform for innovation, creativity, and economic opportunity . . . [and to] . . . advance the development and application of Internet infrastructure, technologies, and open standards to enable an open Internet for all'.

In 2016, ISOC developed a 'Policy Framework for an Open and Trusted Internet'. An important element of this is to improve the security of IoT devices. Encryption, collaborative security, user trust, and Internet restrictions, identity, and privacy are also key concerns within this framework. The goal of improving technical security is one where developing technical standards is particularly important for ISOC through the work of the IETF. One stated priority is to 'contribute to the development of best practices or technical standards resulting in standards that are secure and deployable in global setting . . . by . . . assess[ing] potential deployment issues in the deployment of new standards such as Transport Layer Security (TLS) 1.3, and Domain Name System (DNS) over TLS and HTTPS'.[19] With respect to shaping the Internet's future, ISOC undertakes analysis of key global Internet developments. Its 2019 Global Internet Report highlighted the Internet's potential in terms of its technical evolution and use, but also its problems, arising from a discernible consolidation of the Internet economy.

Unlike the IETF, ISOC membership is organizational. Members are defined as 'any organization that shares an interest and belief in the Society's principles and mission, and meets Organizational Membership dues requirements. Each Organizational Member shall act through its duly authorized representative(s).'[20] Although anyone can join ISOC, its membership is overwhelmingly large corporations (Table 3.3). Organizational members meet at least once per year. Fees range between US$5,000 (professional) to $100,000 (platinum) membership annually providing different levels of representation and weighted voting in 2019.[21] The Internet Society is organized into national Chapters and comprised 110 Chapters in 2019.[22] Chapters are permitted to

[19] https://www.internetsociety.org/key-issues/.
[20] https://www.internetsociety.org/about-internet-society/governance-policies/by-laws/.
[21] https://www.internetsociety.org/about-internet-society/organization-members/membership-levels/.
[22] https://www.internetsociety.org/chapters/.

Table 3.3 Internet Society membership

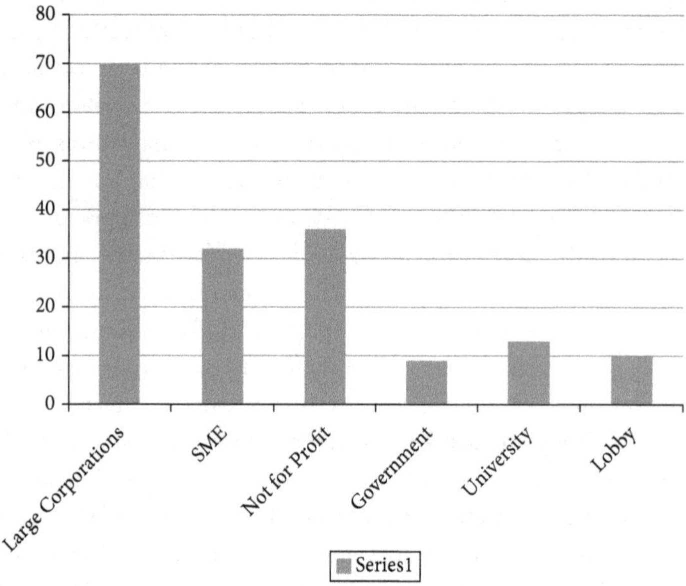

Source: Compiled by Francesca Farmer from W3C membership page. https://www.w3.org/Consortium/Member/List.

pronounce on any matter as long as it can been seen to advance the purposes of ISOC, does not conflict with a position taken by the Board of Trustees, and is unlikely to raise issues of legal or juridical liability.

ISOC meetings contain three types of sessions. Open forum sessions allow physically present and remotely connected participants to ask questions and undertake discussions with the Board. Open to Observers, sessions allow physically present and remotely located listening to occur. Closed Executive sessions, which are restricted to Board members and invited guests, by contrast, deal with matters in which sensitive information is likely to be discussed.

In its 2019 Global Internet Report, ISOC noted the importance of scale in the process of Internet standards development where it 'is not just a source of efficiency, but also a source of power'.[23] On the one hand, the 'open, collaborative, and interoperable Internet is influenced by a small number of

[23] https://future.internetsociety.org/2019/key-features-of-the-consolidation-of-the-internet-economy/.

large companies, and organisational scale and market share play a significant role in the development and deployment of the open technical standards on which the Internet depends' (ISOC, 2019). On the other, it was noted that large organizations could play a positive role in promoting, developing, and testing new standards for the general benefit of the Internet. Interestingly, however, ISOC's 2019 Annual Report, which cited the role of Google in QUIC protocol development which allowed analysis, of usage, experimentation, development, and testing to take place, also noted 'how [Google's] significant browser and content provisioning market share provides it with unique resources that few can match'. A similar point was made by Mathiason when reflecting on the experience of Microsoft's Internet Explorer browser, noting more generally that whilst standards have been a feature of industrial development for centuries, 'the Internet has made them more controversial, since they can affect how well the Internet functions' (Mathiason, 2008:40).

The World Wide Web Consortium (W3C)

The W3C originated from the work of Tim Berners-Lee. He developed HTTP and HTML in 1989 when working at CERN in Switzerland. These protocols underpin the functioning of the World Wide Web. The new graphics interface had wider user appeal to previous file-to-file-based systems such as Gopher, developed at the University of Minnesota, and the File Transfer Protocol (FTP). Berners-Lee had the foresight to release HTTP and HTML as open source protocols to increase use and operability. This followed wide user concern about potential patent development of the web when, before the development of HTTP, Gopher began to demand royalties for commercial use.

In order to prevent fragmentation, Berners-Lee sought to standardize protocols within the IETF. As Russell comments 'Berners-Lee recognized that, absent some sort of institutional effort to coordinate . . . divergent projects, the Web might balkanize into a variety of incompatible standards' (2011:163). Whilst HTML 1.0 was standardized within the IETF, Berners-Lee grew frustrated with the slow speed of decision-making, and established the W3C in 1994 at the Massachusetts Institute of Technology (MIT) initially to develop HTML further as well as Cascading Style Sheets (CSS) which describe webpage layout, and the WWW more generally. HTTP remained at the IETF. The MIT consortium was cloned from older industry consortia. The W3C has since linked up with global 'host institutions' in France, Japan, and China.[24] A key difference between the W3C and the IETF and IEEE is that it operates on an open standard basis.

The W3C was initially financed by DARPA and the European Commission but developed a fee-paying member system over time. By November 2017, the W3C had 471 Members, 150 of which were large corporations, 145 SMEs, 77 non-for-profit, 65 academic members, and 37 from regulatory authorities. Fees ranged from US$2,250 for non-for-profit members with ten or fewer employees to US$77,000 for large private groups in 2019. Individual membership of the W3C is charged on a sliding scale from US$300 for non-affiliated individual members to US$10,000 in 2019 for large corporate representatives, which represents a high financial barrier to entry.

Like ISOC, most membership comes from large corporations but the W3C demonstrates more diverse membership than ISOC which supports the IETF with a greater proportion of small and medium-size enterprises (SMEs), academic and third-sector group membership. The W3C also runs a series of two to three workshops annually, inviting representatives from industry and academics for discussion determined via its 'strategy function'.

The W3C operates differently from the IETF as much of the work is done face to face within working groups which are scheduled one or two times per year. WGs also meet approximately once per week via telephone conferences. Some groups meet more often. Interviews with one group's participants found that a WG met once per week, and in addition, Chairs met on Thursdays and made frequent phone calls to each other. The W3C also holds annual meetings.

Standards are agreed in formal W3C WGs which include paid members, invited experts (invited by the Director, Tim Berners-Lee, and approved by W3C staff), and core personnel. The W3C operates with a relatively small number of twenty-five core staff, twelve of whom are located at MIT in Boston. There are two legal agreements: (i) between the four hosts, which covers employment practices and harmonization rules; and (ii) the membership agreement between each host company and the W3C. In addition, there is an Advisory Board and a Technical Advisory Group (TAG), both elected by the Advisory Committee. The Advisory Committee is composed of representatives from W3C member organizations (see Table 3.4). The Advisory Committee consists of nine to eleven members and has been chaired by Jeff

[24] The regional hubs are the European Research Consortium for Informatics and Mathematics (ERCIM) in France, Keio University in Japan, and Beihang University (in China) https://www.w3.org/Consortium/facts#org.

Table 3.4 W3C membership

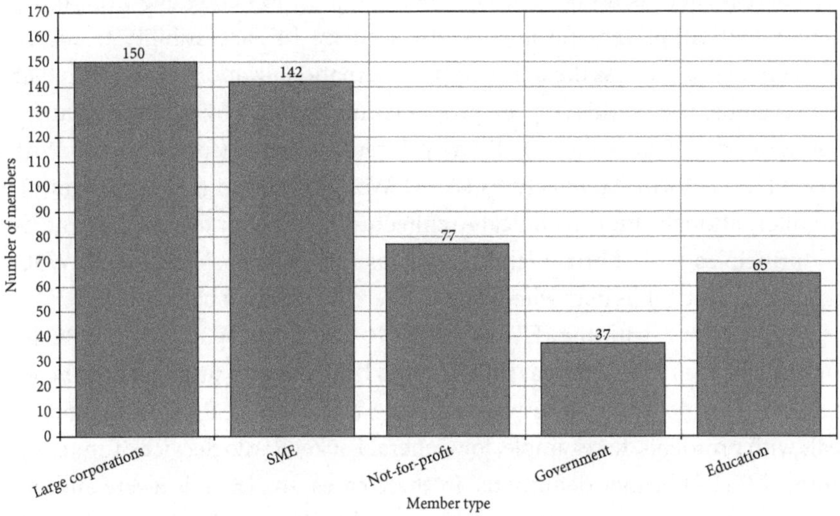

Source: Compiled by Francesca Farmer from W3C membership page. https://www.w3.org/Consortium/Member/List.

Jaffe since 2010.[25] It reviews W3C charters. The TAG was set up in 2001 and consists of eleven people. Chaired by Berners-Lee, it concentrates on technical scrutiny but also discusses wider issues.[26] Examples of these include TAG discussion of the Battery status API and the Google application format. The TAG launched a Self-Review Questionnaire: Security and Privacy in 2019.[27]

WGs permit only one vote per organization. There can be two members from each company present in a WG but only one can vote. The W3C Director then decides whether a proposal goes to final standard, called a W3C Recommendation and is publicized in the Technical Report space. A standard is formalized after two implementations and, if successful, becomes a Recommendation. Before a WG is nominated, there is a review phase wherein the wider W3C community is consulted.

[25] In January 2019, Advisory Board members were: Michael Champion (Microsoft), Jay (Junichi) Kishigami (NTT), Charles McCathie Nevile (ConsenSys), Florian Rivoal (W3C Invited Expert), Natasha Rooney (GSMA), Tzviya Siegman (Wiley), David Singer (Apple), Léonie Watson (TetraLogical), and Hongru (Judy) Zhu (Alibaba).

[26] In January 2019, members of the TAG were Daniel Appelquist (Samsung Electronics) (co-Chair), David Baron (Mozilla Foundation), Hadley Beeman (W3C Invited Expert), Tim Berners-Lee (W3C) (Chair), Alice Boxhall (Google), Kenneth Rohde Christiansen (Intel Corporation), Yves Lafon (W3C) (staff contact), Peter Linss (W3C Invited Expert) (co-Chair), Sangwhan Moon (Odd Concepts), Theresa O'Connor (Apple, Inc.), and Lukasz Olejnik (W3C Invited Expert).

[27] https://www.w3.org/TR/security-privacy-questionnaire/.

There is also a horizontal review[28] with four measures: (i) security, which ensures that there is no cross-leaking from application code and everything is encrypted; (ii) privacy, for information security; (iii) accessibility, to ensure usage for all users across the web; and (iv) internationalization, to establish that there is global use. In this respect, proposals are overseen by the Web Security Interest Group (for security), Privacy Interest Group (for privacy), WAI Accessible Platform Architectures (APA) Working Group (for accessibility), and Internationalization Technical Architecture Group (for technical architecture principles). The horizontal review areas must be considered in all W3C standard-making. It is here where public interest issues can be raised.

There is liaison with the IETF, ICANN, IGF, and the W3C. Most W3C liaison is with the IETF. Whereas the IETF and IEEE develop protocols, the W3C mainly agrees on application programming interfaces (APIs). Sometimes it deals with protocols, for example, for General Packet Radio Services Tunneling Protocol (GTP) header definitions. In these cases, the IETF is aware and informed of W3C developments. There are other overlapping areas with the IETF such as the delivery of on-demand television from CDNs. For example, the BBC is working with both the IETF and the W3C to standardize delivery systems.[29]

In 2019, the W3C had agreed 1,175 Candidate Recommendations.[30] It divided work into seven main areas: web architecture (e.g. OWL, URL, URI, IRI); web design and application (e.g. HTML, CSS, DOM, SVG, Ajax, WCAG and WebApp technologies); the Semantic web (e.g. RDF, SPARQL, OWL, and SKOS); web of services (e.g. HTTP, XML, SOAP, WSDL); browsers and authoring tools, XML technology; and the Web of Devices which includes Mobile Web, Voice Browsing (VoiceXML, PLS, SISR, SRGS, SCXML, and CCXML), Multimodal Access (EMMA, InkML), and Web and TV (e.g. EME) standards. Digital rights groups have been most concerned with EME, as will be discussed in Chapter 5.

The W3C also works on accessibility which is supported in legislation such as in the US General Services Administration's Government-wide IT Accessibility.[31] This is flanked in Europe by ETSI guidelines.[32] The W3C's Web Accessibility Initiative (WAI) was established in 1997. Its key aim was to make

[28] https://www.w3.org/Guide/process/charter.html#horizontal-review.
[29] https://www.bbc.co.uk/rd/projects/dynamic-adaptive-streaming-ip-multicast-dasm.
[30] For a full list of W3C standards, see https://www.w3.org/TR/.
[31] https://www.section508.gov/.
[32] ETSI (2018) Accessibility requirements for ICT products and services. EN 301 549 V2.1.2 (2018-08). https://www.etsi.org/deliver/etsi_en/301500_301599/301549/02.01.02_60/en_301549v020102p.pdf.

the web accessible to users with disabilities. It publishes a series of Content Accessibility Guidelines (WCAG) for developers.[33] The last WCAG 2.1 was published in 2018.

Simcoe notes that the W3C can 'move faster than the IETF because it requires formal membership . . . and places more authority in the hands of a few directors' (2007:270). Most salient work within the W3C is led mainly by browser companies such as Google and Microsoft, manufacturers like Intel, and user companies such as Facebook. Facebook is particularly active within W3C with user case studies. Comcast and Netflix are also important actors, particularly within specifications for television.

Although the W3C runs on a fee basis, it takes its main principles from the IETF including endorsement by W3C and the 'broader community', 'due process', and interoperability which can be found in its Process Document. There are no formal consultation procedures enabling the wider public to participate, but the Process Document[34] states:

> consensus is a core value of W3C. To promote consensus, the W3C process requires Chairs to ensure that groups consider all legitimate views and objections, and endeavour to resolve them, whether these views and objections are expressed by the active participants of the group or by others (e.g. another W3C group, a group in another organization, or the general public).

The W3C has four stages of consultation: the Working Draft (WD) stage; Candidate Recommendation (CR); Proposed Recommendation (PR); and Recommendation (REC). Consensus is considered to take place when a significant number of members support a Recommendation and there is no Formal Objection raised. Formal Objections can be raised directly with the W3C Director who may decide to return a standard to a Group for reconsideration or pass it to final Recommendation.

All W3C standards are published publicly. However, most standards are brought to the W3C when they have been developed and rolled out in the market and the need for interoperability arises. Rarely is it the case that the W3C or even academia develops a standard before the market has developed a proprietary one. Normally once development reaches this stage and a proposal is made, two to three different submissions are made at the same time.

[33] https://www.w3.org/WAI/standards-guidelines/wcag/.
[34] https://www.w3.org/2018/Process-20180201.

Once common ground has been established, a business group or community group is created to develop a CG or FB final draft. Anyone can join a community group. Community groups are held by the W3C largely to encourage wider participation which are seen as incubators for standards.[35] Groups range in size of membership. Community groups present windows of opportunity where civil society can influence proposals and/or alternative technologies can be assessed before a closed WG is established. The number of specifications for open source web platforms is increasing within community groups.

However, only W3C Members or Business Group subscribers can join business groups. Usually, only a small set of a community group joins a WG. Members must sign the contributor agreement and the final report agreement. They also include invited experts. The CG or BG draft is the basis for the establishment of a WG. Members of a WG must act in the capacity of individuals (as Individual Experts) or on behalf of one company or organization only. Participants can be: W3C member representatives; experts appointed by the WG Chair for their expertise; and team representatives. There is only one vote per organization. Thus, if there are two members from the same company within the same WG, there can be only one vote between them. Company representatives are bound to the W3C's Conflict of Interest Policy.

Development within the W3C means that standards are agreed jointly between different market players, meaning a sharing of technology, and often that which has already been patented. Participants must agree to W3C royalty-free licensing before joining a WG. The final product is an open source standard so companies agreeing a standard must publish an open source result, even if patented standards come into the mix. For this purpose, the Advisory Committee establishes a charter. The Advisory Committee also votes on who will be on the Working Group.

After a WG is authorized, a Chair and team contact are assigned to govern progress of the WD. The Resource Description Framework[36] is utilized to produce a relatively complex map for identifying resources and interdependences between one working group and another. The WD is made publicly available and the WG continues work until it produces a Candidate Recommendation. Until 2003, the W3C used to make a last call for comments similar to the IETF. However, since 2003, finalized Candidate Recommendations have been submitted as Proposed Recommendations to the W3C Advisory Committee for approval. The AC votes to accept or

[35] https://www.w3.org/community/groups/.

[36] https://www.w3.org/RDF/.

reject the proposal then the W3C Director declares it as a standard: a W3C Recommendation after which is it is published in the Technical Report (TR). A standard is finalized after two implementations. The W3C differs from the IETF in that there is much more at stake given the high level of membership fees. The IETF process involves many false starts and incomplete work. By contrast, the incentive to finalize a standard within the W3C as quickly as possible is high. The success of proposed standards to final Recommendation (REC) in W3C is therefore greater than in the IETF.

The Organization for the Advancement of Structured Information Standards (OASIS)

OASIS is an international, non-profit, consensus-based open standards body. It was created in 1993 to address interoperability among products supporting the Standard Generalized Markup Language (SGML), the precursor to Extensible Markup Language (XML). It declares that it is 'dedicated to the development, adoption, application, and implementation of structured information standards' and, as a result, its function is to provide an 'open forum to discuss market needs and directions, and to recommend guidelines for product interoperability. The consortium receives, coordinates, develops and disseminates information describing structured information standards and related specifications, methods, recommendations and technologies.'[37] Its main areas of activity in 2019 were security, Cloud computing, IoT, the Smart Grid, content technologies, emergency management, and eGovernment.

OASIS has five strategic goals: provision of an 'effective, efficient, open and transparent environment for the development, coordination and maintenance of high quality standards'; to be broad and international in terms of membership; to 'support all stages of the standards lifecycle'; to develop 'productive relationships with policy setters, analysts, and decision makers affected or potentially affected' by its work; to scan the horizon for areas in need of standardization in order to pre-empt market fragmentation.[38]

Based in Boston, MA, in 2019 OASIS had over 5,000 members from more than 600 organizations and individuals in 100 countries. The organization is led and supervised by a president working under the control of a Board of eleven directors, each of whom serves in an individual capacity (though he

[37] https://www.oasis-open.org/policies-guidelines/bylaws.
[38] https://www.oasis-open.org/org/strategy.

or she must be a member or employee of an OASIS member organization), half of whom are elected annually for a term of two years, and assisted by a Technical Advisory Board. Any organization or individual is permitted to become an OASIS member, meeting once per year and of which there are two categories: voting and non-voting. Similar to the W3C, fees are tiered according to organization type and size ranging from academic contributors (US$1,350 annually) to companies employing more than 500 people (US$54,000 annually) in 2019. Also similar to the W3C, there is a broad constituency of membership in OASIS including commercial players, researchers, users, government agencies, and academic institutions. OASIS is in turn a member of the W3C.

Each member has one vote on each matter considered. Voting can take place by voice or ballot. In the latter case, it is possible to take an action outside a formal meeting through written or email ballot. Here 'any action required or permitted to be taken by the members may be taken without a meeting, if all members shall individually or collectively consent in writing by paper, facsimile, or by email to the action'.[39] The OASIS board can create Member Sections, governed by a steering committee. The latter address 'particular structured information standards or families of standards as designated by its Rules of Procedure' which are created by its members.

OASIS allows member organizations to have as many members as they wish participating in its Committees approved through the organization's Primary Representative. However, committees specify Voting Members and operate on the basis of one member one vote, irrespective of the number of persons from a single organization that are in a committee.

The organization had developed over eighty standards by 2019 and has an IPR policy which contains four modes specifying licensing requirements which are selected by Technical Committee members. For the most part, Non-Assertion or Royalty-Free modes are employed by the organization. OASIS antitrust guidelines state that its policy is 'to comply with all antitrust and competition laws and regulations' and it 'may not . . . in any way restrict competition in OASIS member industries'.[40] These guidelines emphasize that members 'may compete with one another in various lines of business and that it is therefore imperative that they and their representatives act in a manner that does not violate any applicable antitrust or competition laws and regulations'. The organization has established formal relations with the ISO,

[39] https://www.oasis-open.org/policies-guidelines/bylaws.
[40] https://www.oasis-open.org/policies-guidelines/antitrust.

ISO/IEC (International Electrotechnical Commission), JRC (Joint Research Centre of the European Commission), ITU, UN ECE (United Nations Economic Commission for Europe), ANSI (American National Standards Institute), CalConnect, the European Information and Communication Technology Standards Board (ICTSB), Korean National IT Industry Promotion Agency (NIPA), NAESB (North American Energy Standards Board), TM Forum, and the W3C, among others. OASIS also has a community dimension involving the provision of a websites related to its standards (XML.org, OASIS News, and Interoperate IoT).

The Institute of Electrical and Electronics Engineers (IEEE)

The IEEE is another important SDO which develops standards for wireless networking such as Ethernet (IEEE 802.3 set of standards), Bluetooth (IEEE standard 802.15.1), and WiFi (IEEE standard 802.11). It is a global body with more than 426,000 members in 2019 from 160 states that provides global technical fora 'serving the professionals working on emerging and disruptive technologies', in the process using the acquired insights 'to provide governments, NGOs, and other organisations and the public with innovative, practical recommendations to address public policy issues' (IEEE 2015 Strategic Plan). Interviews revealed that, apart from Article 19, civil society groups were not represented within the IEEE.

The IEEE consists of an Assembly and Board of Directors (BoD) selected from its membership. The BoD is crucial in that it has direct responsibility for a raft of boards related to the IEEE's core activities, only one of which is developing technical standards. The latter is headed by the Standards Association Board of Governors and Standards Board. The Standards Board presides over a series of committees which undertake the detailed technical standards making work of the organization. Here, there are as many as thirty-one Societies developing standards. Committees on Radio Frequency (RF) Techniques, Audio and Visual Techniques, Access and Core Networks Standards, Dynamic Spectrum Access, Mobile Communications Networks, Virtualised and Software Defined Networks and Services and Local Area Network/Metropolitan Area Network standards are particularly noteworthy with respect to the subject matter of this book. IEEE Constitutional change needs the support of all IEEE voting members whereas approval by the Board of Directors is sufficient to alter its bylaws (Kanevskaia, 2018).

The Standards Board aims to 'encourage and coordinate the development and revision of IEEE standards; to approve the initiation of IEEE standards projects; and to review them for consensus, due process, openness and balance' as well as to give final approval to standards before publication.[41] It is interesting to note the assertion that its 'membership shall reflect the current standards needs of societies that are developing standards, industry sectors, international interests, and selected organisations (e.g. government, consumer)'. IEEE standards-making is underpinned by the five principles of due process, openness, consensus, balance, and right of appeal. The IEEE SA New Standards Committee, alongside the Standards Review Committee, recommends the approval of new and revised standards requests made to the Standards Board. The Procedures Committee makes recommendations on procedural changes to improve IEEE SA SB efficiency. The Patents Committee has oversight of patents conformity and use and information on IEEE standards. The Audit Committee overseas standards development in the IEEE's Societies, their standards-development entities and the Standards Coordinating Committees of the IEEE SA SB.

The approval by the IEEE Standards Association of a new standard endorses it as 'consistent with good engineering practice' with 'a consensus of representatives from materially affected industries, governments or public interests'.[42] The IEEE SA aims for 'open and innovative' deliberations in the standards-development process and is the only part of the IEEE authorized to represent the organization on external standardization matters. The IEEE SA Board of Governors undertakes to 'establish policy, provide for financial oversight, and conduct standards-related activities within the technological fields of the IEEE'. Within this remit, one of its key responsibilities is to 'establish and maintain a constituency of IEEE-SA members representative of all basic interests' comprising technical societies, industry, government, and the general public. It also has financial oversight and operational management responsibility for IEEE SA.

The IEEE Standards Association specifies membership criteria in IEEE Bylaw 1-403.1. The IEEE levies membership fees, which, for so-called entity members (corporations), varies according to annual revenue. The operations manual outlines that membership of the organization allows voting on 'an unlimited number of proposed IEEE draft standards and on the withdrawal from active status of existing IEEE standards that have reached the Standards

[41] https://standards.ieee.org/about/policies/sa-opman/sect5.html.
[42] https://standards.ieee.org/about/policies/sa-opman/sect1-3.html.

Association ballot stage'. The process of development of a new IEEE standard commences with the proposal of a new work area to an IEEE SA Sponsor. The latter then forms a Study Group, which can be based on an individual member or an entity member, to explore the proposal in more detail. The Group becomes a sub-group of the Sponsor. The Sponsor should consider the potential market demand and technical feasibility of the proposed project; how it would relate to existing standards; whom might lead and participate in its development; its scope and objectives; and the extent to which it could broaden the expertise base of the sponsor. Sponsors are usually IEEE Societies or Committees.

IEEE SA guidelines stipulate that each Study Group should have a Chair and a Secretary, the former of whom circulates a call for participation in the Group's activities. The Study Group should follow the IEEE SA's antitrust policies. If sufficient interest and resources are present, a Project Authorization Request (PAR) is developed in relation to the proposed standard-development project. A project will not be considered to exist until this point, even though in practice members of the working group may have met to work on the PAR. A Study Group can exist for up to six months before a PAR needs to be submitted. Existing working groups may create new PARs.

The IEEE stipulates that in Individual Study Groups members should act on the basis of their individual expertise, whereas in Entity Study Groups it is stated that at least three separate entities should take part. It is recommended that Study Group decision taking occurs consensually, though otherwise on a majority basis when needed. In the latter case, for Individual Study Groups, the recommendation is that at least five people are needed in each vote and that 75% approval is required for the issue at hand to be resolved. In relation to entity-based Study Groups, it is recommended that a minimum of three entities should take part in a vote where 75% approval should be gained for the proposal in question to be accepted. When a Study Group recommends the development of a new standard (after which it is disbanded), it should provide its sponsor with a draft PAR, available intellectual property details, a draft project schedule and budget, and any recommendations regarding external organizational liaison in the development of the project in question. At this point, the process of formal balloting occurs. Following the ballot, the draft standard is submitted to the Review Committee which submits it subsequently to the SB for approval. After review and acceptance the standard is published and can be distributed and purchased.[43]

[43] https://standards.ieee.org/develop/govern.html.

The IEEE provides guidance on the conduct of its standards meetings, where the objective is to secure consensus among government, industry, and the public, and in which any party 'interested in a standards activity or potentially affected by it is permitted to attend any IEEE standards meeting'. It is interesting to note that specific mention of the need to consider any relevant patent issues, pricing, or other anti-trust matters, and any current litigation is made in the guidelines. The New Standards Committee requires PARs sent to it for ratification to have been reviewed by a Sponsor Chair or Liaison. It is usual that Sponsors and Working Group Chairs respond to the New Standards Committee (NesCom) requests within the space of two of its meetings. The normal maximum period to complete a PAR is four years. It is expected that a project which has received PAR approval should open a PAR ballot within eight years.

A key issue for the IEEE, as for all standards bodies, is the role and position of patents in its standards-making processes. The Patent Committee (PatCom) oversees the use of patents and patent information in IEEE standards. In 2015, a key set of changes to IEEE patent policy were approved after a period of amendments to proposals. Under these, IEEE members commit to license patents to users of IEEE standards on fair, reasonable, and non-discriminatory (FRAND) terms. The move was in response to a rise in litigation, which is discussed in Chapter 11.

Conclusion

Understanding the organizational context within which private technical standards for the Internet takes place is a vital part of determining their overall socio-political significance. The philosophical origins of the Internet as an open and growing global communications network pointed to the innate logic of pursuing standardizing behaviour to sit in service of the pursuit of access and ubiquity of use. The informal, communitarian, consultative, and deliberative characteristics of the early-stage Internet were reflected in the organizational features of the IETF. As the Internet grew and became more commercial in its characteristics and strategic significance, these features have come under some strain. The private technical standards-organizational landscape has developed as a result.

As shown, the Internet Society was created to provide the IETF with a degree of financial succour and legal assurance necessary as a result of the removal of public financial support and heightening commercial consequences entailed in standards-making. Subsequent organizational contexts for web standards development, such as the W3C and OASIS have been conscious of the need to take decisions as expeditiously as possible. Long-standing, more hierarchically structured technical standards bodies, notably the IEEE, have become influential in core aspects of Internet standards development.

Yet, the fundamental characteristic of technical standards-making of the need to pursue consensus is still recognizably important. However, the extent to which the pursuit of consensus is underpinned by strategic organizational self-interest over the development of electronic communication for broad human development is an open question and one which underpins much of the content of the remaining chapters of the book. The core concern is that whilst private technical standards development of the kind explored in the case study chapters, largely led by commercial actors or their representatives, can serve the public interest, as echoed by DeNardis, it is important to consider 'the sources of legitimacy for this privatized governance, how the public interest can be reflected, and what is the responsibility of governments, if any, to encourage the promotion of certain types of standards' needs to be considered' (2014:83–84).

4

The Quick UDP Internet Connection (QUIC) and Transport Layer Security 1.3 Standards

Snowden and the Impact on the Encryption Debate in the IETF

Introduction

For twenty years, the Secure Sockets Layer (SSL) and its successor, Transport Layer Security (TLS), provided security for Internet traffic. However, whilst being continuously revised and upgraded, all versions suffered from a series of implementation and security issues, in particular the TLS 1.2 protocol developed in 2008. Following the revelations by Edward Snowden of mass surveillance by the US National Security Agency (NSA) in June 2013, many engineers working on Internet standards vowed to 'encrypt everything' on the Internet in order to ensure the privacy of the end user. Indeed, the Internet Engineering Task Force (IETF) community, asserted in a best practice guide that 'pervasive monitoring is a technical attack that should be mitigated in the design of IETF protocols, where possible' and that 'pervasive monitoring is an attack on the privacy of Internet users and organisations' (Farrell 2014; IETF 2014). Pushback from companies to demands from national security agencies for access manifested itself in standards-developing organizations (SDOs) following Snowden. Cisco, Fortinet, and Juniper, for example, revealed in 2016 that the NSA had successfully targeted their firewalls for years. Contestation within the IETF, however, barred progress and an upgrade to the more secure TLS 1.3 protocol until a final agreement was reached in March 2018 at the IETF meeting in London (IETF, 2018).

It is argued here, in the context of the Multiple Streams (MS) framework, that the Snowden revelations were a major exogenous mechanism—a focusing event in the problem stream—that opened up a window of opportunity

Global Standard-Setting in Internet Governance. Alison Harcourt, George Christou and Seamus Simpson, Oxford University Press (2020).
DOI: 10.1093/law-oso/9780198841524.001.0001

for ideational reinforcement and urgency against pervasive monitoring and a coupling of QUIC and TLS 1.3 to address both latency and security. This is not to say that the IETF did not have a history of engaging with surveillance issues prior to this time. Indeed, the work of DeNardis (2015) shows that the IETF has consistently formed a consensus that does not tolerate technology-driven mass and indiscriminate government surveillance. Rogers and Eden (2017) also demonstrate how the Snowden revelations helped to politicize still further the issue of privacy and security in the IETF (and other SDOs); thus for them, whilst engaging historically with surveillance issues, the Snowden affair prioritized the issue of security, pushing it significantly further up the agenda and this permeated across all relevant working groups in the IETF. The argument here adds to the existing literature to explain how such a focusing event, alongside other internal mechanisms, provided the opportunity for optimizing not just speed but also security with regards to Internet traffic through the parallel development of QUIC and TLS 1.3.

Moreover, this chapter posits that the Snowden revelations provoked a shift in the national mood in the politics stream among citizens of the United States and the European Union towards more privacy and better protection by design for users, and that this was an important factor in reinforcing the ideological shift in the IETF to inject urgency into 'encrypting everything', including through the use of TLS 1.3 with the QUIC protocol. To this end, the explanatory mechanisms with regards to QUIC and TLS 1.3 both have in common the Snowden revelations (as the exogenous mechanism) and, essentially, the coupling of the problem stream with the politics stream, but they also differ with regards to other factors that explain their emergence and contestation and proposed solutions within their respective working groups.

Following the logic of the MSF in Chapter 2, this chapter will identify the underlying problem that led to the QUIC and TLS 1.3 standards being proposed and developed through the IETF. This will include the coupling of the problem with the politics stream and the window of opportunity within which QUIC and TLS 1.3 were discussed. It will then focus on the policy stream and provide an overview and analysis of the progress made in constructing and agreeing the QUIC and TLS 1.3 standards within the IETF life cycle thus highlighting areas of contestation and consensus in relation to both standards in their respective working groups and mailing lists within the IETF. The chapter will conclude with reflections on and the implications of both standards for the security and speed of Internet traffic with regards to adoption and implementation by all the relevant actors.

Coupling the Problem Stream with the Politics Stream and TLS with QUIC

The problem stream provides the background picture to a policy process wherein problems surface which require attention. To this end, adapting the multiple streams (MS) framework, the question asked in relation to QUIC and TLS 1.3 is: what were the generating mechanisms that brought them to the attention of the IETF community? More specifically, can indicators, focusing events, and feedback mechanisms be identified that drew attention to discussion of the QUIC and TLS protocols?

As alluded to earlier, there is no doubt that the Edward Snowden exposure of mass surveillance by the US NSA was the political focussing event that led Internet protocol designers to call for a 'hardening of the Internet' (DeNardis, 2015:72). Evidence suggests that IETF engineers took such attacks personally with regard to the design of their security protocols, and that this was the primary motivating mechanism for the movement to 'encrypting everything'. The more general political shift among citizens and indeed privacy advocates towards having better legal protection with regards to privacy and data protection in the European Union and also the United States meant that there was also a change in broader 'national' mood against mass surveillance towards control and consent (even though there existed more resistance and constraint in the United States to achieving this; see Chapter 7, for example).

Indeed, this was confirmed in the interviews conducted at the IETF in terms of a general sense and feeling that 'knowledge of pervasive global monitoring . . . Snowden and basically a more widely known understanding of the surveillance apparatus helped push [TLS 1.3] forward' (Interview, TLS WG member, IETF Prague, July 2017; Interviews IETF Prague, 2017, IETF London, March 2018; virtual interviews, 2017, 2018). It was clear that the Snowden affair not only provided impetus but also an opportunity to revise the previous underlying encryption philosophy towards a more secure forward secrecy encryption. Moreover, the publication of RFC 7258 in May 2014 (IETF 2014) stating that '[p]ervasive monitoring is a technical attack that should be mitigated in the design of IETF protocols, where possible' put in place a process that would affect every strand of work that the IETF carried out with regards to the development of standards (Rogers and Eden, 2017:811). The norm of pervasive monitoring and technical mitigation now had to be considered by all IETF working groups (WGs) alongside the security properties of any standards being developed.

However, there were also other generating mechanisms that meant the TLS and QUIC protocols were brought to the attention of the IETF community for revision and development with various implications for the role of civil society actors in terms of their influence and input. Snowden, effectively, provided further impetus in 2013—symbolically and ideationally within the IETF and other SDOs—for creating protocols to protect against future assaults on individual privacy through redesigning protocols that ensured that security was front and centre when it came to communications on the World Wide Web (WWW). QUIC and TLS, however, appeared on the IETF agenda prior to this. How certain indicators and feedback mechanisms—formal and informal—as well as the actions of policy entrepreneurs resulted in these protocols becoming prominent on the IETF agenda have to, therefore, be taken into account.

The Transport Layer Security Protocol (TLS 1.3)

The TLS protocol was initially developed by Netscape Communications and named the Secure Sockets Layer (SSL). SSL officially came to the IETF in the mid-1990s, where in 1999 the first iteration—TLS 1.0—was released. This was to be the first of several versions that were developed prior to TLS 1.3, but important here is that there have been a series of triggers since 1999 that led to its revision and the subsequent release of TLS 1.1 (2006) and TLS 1.2 (2008). To this end, evidence suggests that, in addition to Snowden as the main exogenous mechanism, three other central internal mechanisms were at play that led to TLS 1.3 becoming more prominent on the IETF agenda and within the TLS WG at different times after 2008.

First was the widening of the actors involved in the process and indeed the increased legitimacy, credibility, and recognition of these actors in the TLS protocol testing, review, and feedback process. There was an increase in academic work on formal security models for protocols that demonstrated the problems with TLS 1.2 (when TLS 1.3 was initially proposed), and subsequently a push to involve and integrate the academic community to ensure the cryptography in TLS 1.3 was as robust as possible. Paterson and van der Merwe (2016:160–1) argue that the increased involvement of the research community since 2008 had implications for the revision and discussion of TLS in the IETF in several ways: first, in terms of the weaknesses identified by the research community acting as a key feedback mechanism for the IETF in their redrafting of the next version of the protocol, TLS 1.3, in 2014; and second, in moving from a largely reactive culture within the TLS Working Group in the IETF to a proactive culture of redesign.

The work done by many academics (mathematical cryptographers) is closely related to the second aspect that acted as a mechanism for seeking to revise and improve TLS; that is, the number of vulnerabilities identified in various TLS versions—in particular TLS 1.2—following a series of high-profile attacks such as the Renegotiation attack, BEAST, and CRIME. Indeed, such attacks are well-documented by the IETF (RFC 7457[1]), but also by prominent academics that sought to highlight the security strengths and limitations of the TLS protocol. Thus, some have argued that the number of academic research papers increased exponentially and that 'what started as a trickle of papers has . . .become a flood' (Gajek, Manulis, Pereira, Sadeghi, and Schwenket, 2008; Morrisey, Smart, and Warinschi, 2008; Maurer and Tackmann, 2010; Krawczyk and Wee, 2013; Paterson and van der Merwe, 2016:160). This, alongside improvements in technology—both in terms of more mature analytical tools and cryptographic ciphers (crypto algorithms that work together to secure traffic)—alongside a more open dialogue with the academic research community meant there was arguably a shift in the TLS standardization process, with Snowden's actions adding further urgency to the need to make TLS more secure under the ideational banner of 'encrypting everything'. As surmised in the albeit sparse and formative literature that has been published on TLS, 'Prior to the standardisation of TLS 1.3 . . . protocol analysis tools were not mature enough, the research community's involvement . . . was minimal, and until the first wave of attacks in 2009–2012, attacks on TLS were not considered to be of enough practical import to warrant making changes with urgency' (Paterson and van der Merwe, 2016:161).

A final mechanism that must also be considered can be characterized as commercial: the link between increased security and the need to ensure that this did not come at the cost of improving performance. Indeed, it was a key consideration that also meant TLS 1.3 was coupled with the QUIC protocol. It was recognized that in order to make TLS 1.3 a more attractive proposition for adoption to those that did not consider increased security a priority, speed of connection was vital, and thus it became a key priority for those working in the formalized IETF TLS WG but also for those experts involved in the informal deliberations and processes that complemented this (Interview, TLS WG member, IETF Prague, July 2017; Interviews IETF Prague, 2017, IETF London, March 2018; virtual interviews, 2017, 2018).

[1] https://tools.ietf.org/html/rfc7457.

Quick User Datagram Protocol (UDP) Internet Connection (QUIC)

QUIC appeared and became prominent on the IETF agenda for qualitatively different reasons than TLS even though the two protocols become interlinked within the IETF process. It was external (market) factors and impetus that meant a (pre) version of the QUIC protocol was proposed by Jim Roskind, an engineer at Google (Roskind, 2012).[2] Google began work and released QUIC (the Google version known as G-QUIC) in 2013 before it was taken to the IETF for further development in 2015 (the QUIC WG was established in 2016). In this sense, Google can be identified in the SDO context as the initial policy entrepreneur that sought to design and roll-out the QUIC protocol within its own platforms prior to seeking the development of this protocol at the IETF. The question from an MS framework perspective is why Google wanted to take this to the IETF and why the IETF was, equally, willing to develop the protocol. What were the main mechanisms that led Google to develop the QUIC protocol independently and then propose its further development through the IETF—in particular as in 2014 those developing G-QUIC had ruled out IETF standardization (Ermert, 2017)? Several mechanisms can be identified that include commercial- (performance) related motivations, regulatory steer, and technical issues of latency and ossification of the Transport Control Protocol (TCP) through which Internet traffic passes.

First, and in terms of indicators and a feedback mechanism, studies were undertaken that analysed the impact of latency on the revenue of certain companies. Several early evaluations undertaken by Google in 2008—and subsequently other major Internet players—demonstrated that the speed at which users were able to open web pages had an impact on satisfaction, experience, and more importantly, revenue, as slower speed meant that people navigated away from web pages or cancelled sessions (thus, users did not complete transactions or there was a loss of advertising revenue). For example, via several performance case studies, Marissa Mayer (Vice President of Search Product and User Experience at Google) demonstrated that if there was an increase

[2] The original design and rationale behind this protocol drew significantly from work by Jim Roskind (EARLY-DESIGN). The contributors to the pre-IETF QUIC project at Google are: Britt Cyr, Jeremy Dorfman, Ryan Hamilton, Jana Iyengar, Fedor Kouranov, Charles Krasic, Jo Kulik, Adam Langley, Jim Roskind, Robbie Shade, Satyam Shekhar, Cherie Shi, Ian Swett, Raman Tenneti, Victor Vasiliev, Antonio Vicente, Patrik Westin, Alyssa Wilk, Dale Worley, Fan Yang, Dan Zhang, and Daniel Ziegler. See: https://datatracker.ietf.org/doc/draft-ietf-quic-transport/?include_text=1.

in the number of search results per page from 10 to 30, with a corresponding increase in page load times from 400 milliseconds to 900 milliseconds, this resulted in a 25% drop off in first result page searches. Adding the checkout icon (a shopping cart) to search results made the page 2% slower with a corresponding 2% drop in searches/user.

Furthermore, in relation to Google Maps, it was illustrated that image optimizations made the page two to three times faster, with significant increase in user interaction with the site. Other studies, such as that from Shopzilla, for instance, demonstrated more clearly the link between performance and revenue, showing that an increase in speed of 5 seconds (from 7 to 2 seconds) resulted in an increase in page views of 25%, and an increase in revenue of 7–12%.[3] A proliferation of such studies on latency demonstrated and drew further attention to the issue of performance and its link to revenue, acting as a major incentive to the improve efficiency for the user.

Performance, however, was not the only indicator that led to Google thinking further about addressing the issue of latency through QUIC. Another issue related to what engineers call the ossification of the protocol, mainly by middleboxes (e.g. firewalls, inspection software, etc.); in simple terms, this refers to anything that filters or attempts to modify network traffic on the Internet. Whilst the primary aim of QUIC was to address latency, the post-Snowden ideological turn (or at least increase in urgency) in the IETF also brought to the fore the secondary aim; preventing interference and limiting ossification by middleboxes through encrypting everything. Thus, there was a coupling of the latency issue with the issue of improving security with QUIC. G-QUIC was developed so that it could implement reliable, high-performance delivery with congestion control at the application layer (using UDP as the transport layer; see Figure 4.1), supported by all Google services and Google's Chrome browser.[4] Indeed, from 2016 it is estimated that QUIC is used in more than 85% of Chrome requests to Google Servers and that it represents approximately 7% of all Internet traffic due to the popularity of Google services such as YouTube. Initial performance results from Google show significant gains compared to TCP for the slowest 1% of connections and for video streaming. More precisely, according to a Google evaluation, there is a 3% page load time

[3] For a summary of other studies see: https://perspectives.mvdirona.com/2009/10/the-cost-of-latency/). See also https://www.internetsociety.org/blog/2014/11/reducing-internet-latency-the-long-term-challenge-of-making-the-internet-faster/.

[4] Note, however, that network operators only passed recognized traffic, which was a problem when G-QUIC was deployed using UDP on other networks (around 10% failure).

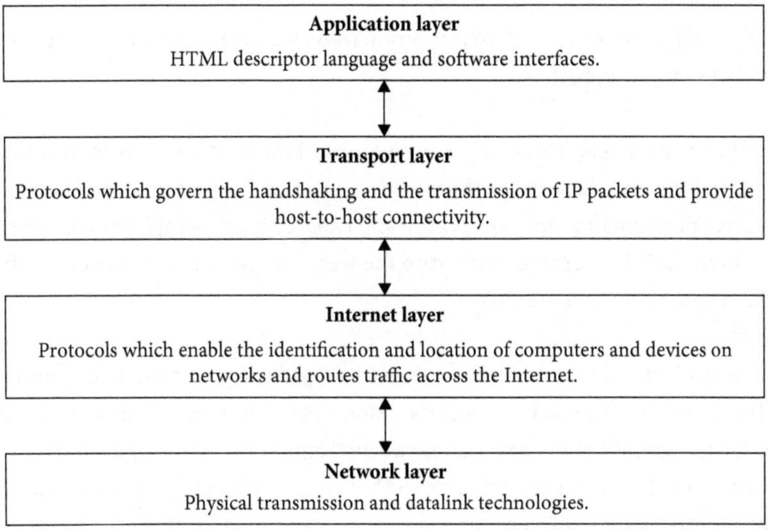

Figure 4.1 Network layers
Source: (Adapted from) Contreras, 2016:856.

improvement on Google Search, an 8% latency reduction on Google search, and an 18% reduction in buffer time on YouTube (Feldman, 2017; Rüth, Poese, Dietzel, and Hohlfield, 2018). Google also developed its own cryptography for QUIC, SPDY (this is not an acronym); a network protocol designed to manipulate Hypertext Transfer Protocol (HTTP) for the specific purposes of reducing latency and increasing web security. Whilst SPDY was used as a starting point for what became http2, one of the prerequisites—one might argue a key condition—for the IETF developing the QUIC protocol (which has become http3, Cimpanu, 2018; Pardue, 2019) was that it would be used with TLS 1.3—especially given what were perceived to be the advantages of this in relation to SPDY with regards to encryption and security (Interview, IETF Prague, 2017). As one QUIC WG member noted, 'the big condition was that we replace this Google security protocol [SPDY] with TLS 1.3 . . . and engineer that. . . . TLS 1.3 was a huge effort [and] compared to TLS 1.2, it's much better security . . . [and] the new version, TLS 1.3 . . . is functionally equivalent or better than what [Google] had' (Interview, IETF Prague, 2017).

Given that Google had engineered QUIC and started to implement it, why was the company motivated to take this to the IETF for further development? The first reason relates to the market or commercial (exogenous) mechanism; Google wanted to take QUIC to the IETF for development so that it would be attractive to other network providers. From a Google perspective, Google ads and Google services are on many other websites and there are also browsers that visit Google websites,

> so they want to get it into other browsers and on to other servers to get the full performance benefit for the entire internet. If the web is not investing in captive performance and isn't secure and capable it will atrophy as a platform so it's in their interests to make sure the web is as good as it possibly can be. (Interview, IETF Prague, 2017)

A second important reason is related to regulatory compliance; that is, industry steer in Internet standards. The Payment Card Industry Security Standards Council (PCI SSC) changed its regulation with regards to the use of credit card handling compliance so that companies had to switch from SSL/early TLS to using later versions of TLS by 30 June 2018 in order to meet the PCI Data Security Standard (PCI DSS) for safeguarding payment data (Gray 2017). Thus, use of QUIC for handling credit card protection as originally conceived by Google would have not been allowed, but this hurdle would be eliminated within the IETF given the proposed use of TLS 1.3 with QUIC.[5]

A final (internal) and interrelated mechanism, however, is a sense of 'doing the right thing' among engineers that construct protocols and seek to work in the public interest in order to ensure that the Internet works as well and as securely for all users. One privacy and rights activist that participated in the QUIC and TLS WGs summarized these twin motivations very well when he stated that 'we can characterise them as commercial advantage but some of them also have real nice consequences for some of the goals of civil society' (virtual interview, 2018). Indeed, even though certain digital rights activists were quite cynical about the motivations of Google, they did acknowledge—confirming again, a distinct culture among certain engineers of the Internet—that there are 'some individuals at Google who are legitimately trying to do the right thing . . . trying to help people communicate' (virtual interview, 2018). From an IETF perspective then, it was not simply a case of 'taking G-QUIC and just putting a nice ribbon around it' (Interview, IETF Prague, July 2017). Rather, the IETF wanted to ensure that they engineered their own QUIC protocol in what they considered to be the right way; that is, that all choices were considered that would lead to both greater efficiency and better security, in particular after Snowden (i.e. encrypt everything). Moreover, given the

[5] See, for example, comments by Roskind: https://groups.google.com/a/chromium.org/forum/#!topic/proto-quic/S5KkNjK4EUQ.

weaknesses identified in previous TLS versions and the issue of ossification, the IETF wanted to ensure any changes made to QUIC in the IETF reflected the general mood to ensure that there were no elements of it (in particular TLS 1.3) that could be 'broken' further down the line.

Policy

The TLS 1.3 and QUIC working groups have deliberated over the specific technical requirements of the protocols, with contestation visible between the different actors, in particular on the meaning of 'encrypting' everything and its implications for actors' preferences and interests. To this end, it can be argued that similar patterns of contestation emerged with regards to QUIC and TLS 1.3 in relation to encryption—albeit in relation to different layers of the Internet stack—with a plethora of proposals debated in an effort to reach rough consensus. Ironically, the most visible contestation related to TLS 1.3, even though rough consensus was reached on this protocol in March 2018. In line with the MS framework, this section will focus on the 'policy stream' and identify the main proposals from the relevant actors in relation to both protocols, the issues of contestation, and which mechanisms were at play in moving the protocols forward towards consensus (in the case of TLS 1.3) or further deliberation in the case of QUIC (now http3).

The Transport Layer Security Protocol (TLS 1.3)

Some of the major Internet industry players have been the main authors of the TLS 1.3[6] protocol: Google, Microsoft, Akamai, Mozilla, Apple, Amazon, Cloudfare, and Facebook. In terms of participation, in particular within the IETF TLS WG meetings, however, the range of actors has been diverse—from US NSA representatives and US Bank, to civil society organizations such as the European Frontiers Foundation (EFF), American Civil Liberties Union (ACLU), Article 19, and the Centre for Democracy and Technology (CDT)—that have, whilst numerically outnumbered by corporate stakeholders, made consistent contributions in support of their goals (privacy, security, transparency, etc.). Indeed, contestation has reflected the politics of different groups within the context of an overarching aim, written into the IETF in the form of

[6] https://datatracker.ietf.org/doc/draft-ietf-tls-tls13.

RFC 7258 (IETF 2014) since Snowden in particular but also RFC 2804 prior to this on Wiretapping (IETF 2000), to ensure that that the fourth iteration of TLS was as secure as possible. As one prominent member of the TLS WG stated: 'especially since the Snowden stuff, I think that sort of radicalised the engineering community who saw their protocols being used against them and figured never again' (Interview, IETF Prague, 2017).

To this end, much of the debate around the design of the TLS 1.3 protocol has been the type of cryptography that should be usedand, in particular, discussions regarding how cryptographic weaknesses could be reduced based on lessons learnt from previous iterations. Incorporating the property of 'forward secrecy' has been at the centre of this. As explained by one engineer involved in the TLS 1.3 WG:

> concrete philosophies that TLS 1.2 had previously encompassed was the ability to have protocol and cipher agility... [so that] you can use this protocol but you can plug in basically any cryptography that you want and you can mix and match and the protocol itself lets you negotiate between client and server which cipher you're going to use. (Interview, IETF Prague, 2017)

The issue with this was that it involved use of cryptographic ciphers (algorithms that work together to secure Internet traffic) that were weak and did not provide forward secrecy; that is, they were vulnerable to attack as shown in many studies (see earlier) because adversaries could figure out what previous keys had been used and decrypt conversations. Forward secrecy, on the other hand, used in a cryptographic system, uses one random key in any given session to complete a key agreement, and the session key is then destroyed once the session ends to prevent an adversary from decrypting traffic afterwards. In other words, it takes away the weakness of non-forward secret cipher suites used in TLS 1.2.[7]

When the property of forward secrecy was proposed on the TLS mailing list—and indeed discussed at several academic workshops (Farrell 2017)there was a broad consensus that this was a desirable move towards creating a more secure protocol (as opposed to using static Diffie–Hellman keys with non-forward secret properties). To this end, in the early stages of the TLS 1.3 design process, 'Snowden and basically a more widely known understanding of the surveillance apparatus helped push one of the philosophies forward which

[7] In basic cryptographic terms it boils down to having a static key over time and with forward secret connections every new connection has a brand-new key.

is forward secrecy' (Interview, IETF Prague, 2017). Despite this, however, the contestation visible in terms of the discussions on the mailing list and in the TLS WG meetings and other observers confirmed that such a consensus was not underpinned by concrete commitment among all actors to the changes necessary in the network architecture and, more concretely, in terms of investment in new technologies and systems. More specifically, it became apparent in the latter stages of the TLS 1.3 design process that 'some enterprise network operators began to realize that this would reduce their ability to inspect traffic in order to troubleshoot problems in their networks' (Checkoway, 2017; [TLS] Industry Concerns about TLS 1.3). In essence, these network operators argued for mechanisms that would allow them to support their internal network needs regarding fraud monitoring, troubleshooting, intrusion detection, and protection, etc., and to this end, several proposals were put forth that called for continued use of a static key, such as that on static Diffie–Hellman keys proposed by Matthew Green and colleagues[8] at IETF Prague, 2017 and by Ralph Housley and Ralph Droms[9] at IETF London, 2018, to allow them to do this.

The debate, however, has been more nuanced than simply forward secrecy advocates versus static non-forward secrecy advocates (the use of static keys or use of older cypher suites that are non-forward secret). As noted by Checkoway (2017) in his observations of the evolving TLS 1.3 design process, both within the WG meetings and on the mailing list, three perspectives can be discerned: that of the enterprise networks, that of the forward secrecy grouping, and, finally, a more pragmatic group that has sought to propose compromises that sit between the other two camps. In terms of the first perspective, network operators argued that their ability to decrypt traffic within their network was essential in order to be able to manage their network effectively, and that 'deprecation of the RSA key exchange in TLS 1.3 will cause significant problems', in particular for financial services companies ([TLS] Industry Concerns about TLS 1.3). It was not only financial services companies that advocated this perspective, however. As one participant acknowledged, it was also 'Distributed Denial of Service mitigation providers and people who sell load balancers which is a piece of hardware that will send a message to one part of your data centre rather than another to help you keep traffic where it should be' (Interview, IETF Prague, 2017). This, however, has been challenged vigorously within the IETF, both at the TLS WG meetings and on the mailing lists. For those that support forward secrecy in particular

[8] https://tools.ietf.org/html/draft-green-tls-static-dh-in-tls13-00.
[9] https://tools.ietf.org/html/draft-rhrd-tls-tls13-visibility-00.

(and thus use of ephemeral rather static Diffie–Hellman keys) there are several major concerns.

The first relates to the implications of being able to decrypt traffic even if only within enterprise networks. To this end, it is argued by some (and heavily contested by others) that this sort of proposal enables wiretapping as 'there is no way . . . to constrain the use of this scheme . . . to such (part of) networks' (Farrell, 2017). In other words, it is not technically possible to constrain such activity within a particular network or data centre. Second, it is argued that weak design (weak cryptography) that would essentially lead to breaking TLS would mean not learning the lessons of the past given the history of TLS (and SSL). Thus, many individuals voiced concerns that designing a weaker TLS 1.3 version would have a detrimental impact in the future. Moreover, it was asserted that 'the IETF is not bad at making security technology better, we've never tried to make it worse and we would fail. Anytime you introduce these kinds of weaknesses they come back to bite you' (Interviews, IETF Prague, 2017). A good example of this are the export cipher suites with cryptographically weak algorithms which were at a certain point in time the only cipher suites that could be legally exported from the United States. As Checkoway points out: '[t]wo decades after the use of export ciphersuites should have ended, researchers showed how to abuse support for these deprecated algorithms in modern TLS libraries to man-in-the-middle TLS connections' (2017). From a forward secret perspective then—which includes those civil society actors that seek to ensure privacy is a primary consideration in any protocol design—a weak design of TLS 1.3 should not be countenanced no matter what the cost for network enterprises; from a network enterprise perspective, a major reason for their resistance was that products would 'need to be replaced or substantially redesigned at significant cost and scalability' ([TLS] Industry Concerns about TLS 1.3).

Finally, in terms of the third pragmatic viewpoint there were those that sought to chart a compromise between forward secrecy advocates and those enterprise network operators that wanted to maintain some decryption that they perceived as essentially to their operational capability. The argument underpinning much of this was that those that really wanted to decrypt to maintain operational capability would end up doing this in implementation; so from this perspective it was deemed more appropriate to find a compromise within the IETF than to end up with a patchwork of weak cryptographic implementations among key industry players that found the forward secrecy element of TLS 1.3 problematic. To this end, several proposals emerged that suggested trade-offs between network visibility and forward secrecy including client opt-in (where the client is aware of interception; Green, 2016; Housey and Droms, 2017) but also use of alternative encryption protocols being used within datacentres (Moriarty, 2017).

At the London IETF meeting in March 2018, two questions were posed to wrap up the TLS WG meeting relating to the trade-off option presented by Housey and Droms (2017): (i) do you support adoption of draft rhrd-tls-tls13-visibility, or (ii) do you oppose adoption of draft rhrd-tls-tls13-visibility? The hum (i.e. to indicate rough consensus) on these questions revealed that there was no support for compromising on the TLS 1.3 protocol through providing visibility on any part of the network even though many in the room expressed concerns about various technical issues that had not been addressed. TLS 1.3 was thus agreed on 20 March 2018, and the central issue for the future is that of implementation, which 'will require people to put in some effort to make it work properly'. At the time of writing, whilst the pragmatists did not win out in the IETF, it is most likely that a practical technical interim solution will be found to make TLS 1.3 operational whilst accommodating the needs of network enterprises. Indeed, one of these solutions is to make TLS 1.3 look like TLS 1.2, which means that 'some of the worst aspects of TLS 1.2 . . . that hackers could trick the system into reverting to an old insecure ciphersuite—are plugged without customers having to take a large-scale upgrade to existing systems' (Interview, IETF Prague, 2017; Chirgwin, 2018). Moreover, there is an acknowledgement that there is a learning curve and that 'in the long term people will come around and understand that the high level of security is going to do them a lot of good' (Interview, IETF London, 2018).

Quick User Datagram Protocol (UDP) Internet Connection (QUIC)

Similar to TLS 1.3, the QUIC WG in the IETF is well represented by the major industry players, with the same civil society voices contributing both formally and informally to discussions since the WGs inception in 2016 (https://github.com/quicwg). The debates on QUIC within the IETF QUIC WG meetings but also on the mailing list, it can be argued, were driven by similar concerns in relation to encryption and visibility of data or metadata that would be lost by network operators, as well as the need to eliminate interference from middleboxes. The use of encryption in QUIC, however, is different to that in TLS 1.3. With regard to the latter, the aim is to ensure the privacy and integrity of the payload so that third parties cannot interfere with the web page you are loading over https. QUIC, however, takes this further in that it encrypts not just the payload but also the control information (it encrypts the transport headers), which for some is essential for privacy and security but for others, brings no benefit to either (Snellman, 2016; Bouzas 2016). In essence, the debate on QUIC touched on several interrelated issues for network operators and indeed for certain mobile network operators (MNO) in particular. Not all MNOs took the same position on QUIC. Indeed, positions varied from those that questioned the need for it to those that viewed it positively but nevertheless argued that there needed to be a certain level of exposure of data for them to be able to ensure they could carry out quality assurance and management functions (Sarker, 2018). From a civil society perspective, the central concerns in the discussions over QUIC related to issues of privacy and resistance to exposing any metadata within QUIC for network measurement or planning.

A central issue that was raised was that of MNOs not being able to see the traffic flowing through their networks and, in particular, the redundancy of traditional optimization tools with the use of QUIC. The arguments within the QUIC WG meetings and the discussions on the mailing list, therefore, centred on how some element of measurement could be maintained within QUIC; that is, how data or metadata could be measured, albeit passively, in order to allow network operators to carry out key operational tasks. To explain further: using TCP made a lot of information visible to the path; that is, network operators could see deeply into the IP and transport header packets to 'perform traffic engineering functions, packet interception and forced proxy caching'. Moreover, they could reach 'deep into TCP control fields and manipulate these values to modify session flow rates' (Huston, 2018). Essentially, network operators became used to and saw as essential to their service the ability to do this.

Indeed, whilst many network operators, in principle, see QUIC as a positive move forward with regards to protocol design, they nevertheless also argue that a solution should be found that will allow operators to at least passively measure, for example, round trip times (RTT) or latency on the path, 'without exposing information that would lead to any significant privacy implications' (Interview, IETF Prague, July 2017). Whilst privacy advocates do not agree with this position, a central debate in the IETF has been about the how and how much information should be exposed, and it is here, in the detail, that the main disagreements lie. There are network operators that argue that privacy issues are minimal even within TCP (QUIC WG meeting, IETF Prague 2017), but the more pragmatic network operators have sought to engage in a discussion on a minimum set of information that could be exposed which would allow them to measure latency and make certain inferences on packet loss or reordering. From this perspective, the position was that 'operators needed to be able to manage the network whilst accepting that Internet traffic would be encrypted' (Smith, 2017. From a privacy perspective, however, there has been pushback on even minimal exposure, and this has not just come from civil society representatives. Indeed, the debates around QUIC were well exemplified by discussions in the QUIC WG meeting in Prague in July 2017, where various options were discussed from doing nothing (i.e. not even passive measurement of RTT) to exposing one or more 'bits' of information. The network and, in particular, mobile network operators that were in favour of the latter options argued that these were the most basic for network measurement and planning that would allow for traffic optimization as they exposed 'such an insignificant bit of the entire sequence numbers that it reduced issues of fingerprinting or linkability or things like this' (Interview, IETF Prague, July 2017). Moreover, it was asserted that such options represented an acceptable trade-off between any future privacy risks and the need of network operators and they would ensure network operator buy-in which would be essential to prevent operators from blocking QUIC and defaulting to TCP if they felt this was necessary for their day-to-day operations. This worst-case scenario would, it was pointed out, have far more severe privacy implications than actually finding a compromise and balance on QUIC that would lead to broader implementation.

Resistance came from those in the QUIC WG—most consistently from civil society voices but not exclusively—that believe any exposure of 'bits' of information would be dangerous. The reasoning given for this position is twofold: (i) the potential consequences of exposing any information or allowing any form of passive measurement in the long term are not known, in particular given that any such exposure in the past (e.g. port numbers), provide evidence of the damage network operators have done and how they have gone further than necessary (e.g. deep packet inspection), thus compromising the privacy and security of the end user. (ii) Even with the basic trade-off and minimal exposure, visibility of response times enables geolocation to a set of approximations. Moreover, it was also argued that any information about timing in terms of endpoints would allow inference in relation to a range of issues that again, compromise privacy. Others also pointed to the link between privacy problems and ossification and the need to avoid ossification and leakage, one of the central aims of both QUIC and indeed TLS 1.3. From this perspective then, a precautionary approach was needed that minimized what can be done to networks, ensured end-to-end privacy, and, importantly, did not lead to a scenario in the near future where QUIC 2.0 needed to be developed due to any errors exposed. Importantly, it was argued that the technical arguments for why network managers needed to be able to measure data were not convincing, with one activist asserting:

> I think they don't need it and frankly if they spent the money that they're spending on deep network inspection instead on increased capacity and faster backplane they could just pass more traffic anyway and it's probably at least as good and simpler to manage. Some of these folks are also network equipment operators or vendors who believe that their positioning in the market is based on differential products. So you end up with some network equipment vendors who have, in some sense, a stake in the network being analysable. (Virtual Interview, QUIC WG participant, October 2017)

Such vested interests are critical in the development and agreement of any protocol in the IETF; with QUIC in particular, certain QUIC WG participants pointed to how those vendors and services with vested interests 'stacked the room' and brought people in 'to show up for one meeting and hum very loudly' in order to ensure their interests were represented (Interviews, QUIC WG participants, EITF Prague 2017/London 2018). Whilst London IETF WG discussions in March 2018 focused on further technical issues around both RTT and the architecture for QUIC, and this has continued on the list and subsequent IETF QUIC WG meetings, observations are that once these are ironed out, QUIC will arrive at a rough consensus sooner rather than later.

Conclusion

This chapter outlined the emergence, contestation, and potential solutions related to two IETF standards, TLS 1.3 and QUIC—, hat have sought to improve the speed at which end users can use the Internet, as well as protecting the privacy and confidentiality of end users and their data through, essentially, end-to-end encryption. The QUIC standard, initially developed by Google (as the policy entrepreneur) and passed to the IETF, whilst offering multiple benefits, was contested by network operators—in particular MNOs—that argued it would affect their day-to-day ability to locate and efficiently resolve problems and improve security and performance within networks through the use of middleboxes.

Privacy advocates, on the other hand, argued that there was no option but to encrypt to prevent manipulation and ensure the privacy of users going forward. This battle between cryptographers and network managers is not easily resolvable and will require compromise for a suitable solution that satisfies both parties to be found. The state of play at the time of writing suggests that the 'devil is in the technical detail' and that discussions towards achievement of a rough consensus is only a matter of time now that the work of the QUIC WG has effectively been transferred to that of the http3 WG. The TLS 1.3 protocol witnessed similar arguments, albeit related not to the application layer but rather to the transport layer, with rough consensus reached on the protocol underpinned by the key principle of forward secrecy. Here, the challenge will be in the technical implementation of the protocol, in particular in relation to the existing architecture.

This chapter has demonstrated that the Snowden revelations were a major exogenous mechanism—a focusing event in the problem stream—that opened a window of opportunity for ideational reinforcement and urgency against pervasive monitoring and a coupling of QUIC and TLS 1.3 to address both latency and security. It has also importantly shown that other mechanisms can impact on the contestation and development of QUIC and TLS, from exogenous market or commercial mechanisms to internal collective and individual (cultural) mechanisms. In terms of QUIC, it was clear that the IETF community was only willing to adopt and standardize it according to its own 'internal' sense of ensuring the engineering was right—that is, that it would offer the best security as well as efficiency. With regards to TLS 1.3, where work and reflection began much earlier than Snowden, feedback from a broader community of academic mathematical cryptographers was important in terms of moving the TLS WG from a reactive to proactive mode of redesign. Snowden then, was no doubt the primary factor that injected the sense of urgency and, indeed, galvanized a more radical stance among engineers in terms of their determination to 'encrypt everything'. It is important, however, to acknowledge that other mechanisms, before and after Snowden, allow a deeper understanding of how these two standards have evolved, been coupled and contested, and how they will be implemented in the future.

5

Political Drift and Forum Shifts

The Case of Browser Development

Policy drift refers to situations in collective decision-making where stasis in decision-making exists internally to an organization when the external context changes (Hacker, 2005). In this situation, a given organization suffers from lack of reform or adjusts poorly to new environments. It often results in what Hacker referred to as 'layering' where new rules are introduced on top of or in addition to old rules. Layering is seen in this chapter as a mechanism which has led to windows of opportunity for change within the context of the multiple streams (MS) framework. According to the public policy literature, layering occurs where there is a low level of regulatory oversight or implementation is only voluntary. New schemes are formed even though the older ones continue to exist. This leads to the creation of new internal constituencies and boundaries (Béland, 2013; Falleti and Lynch, 2009:8). This occurred with the development of mark-up languages within the World Wide Web consortium (W3C) and The Web Hypertext Application Technology Working Group (WHATWG).

The first half of this chapter shows that stasis in development of Hypertext Markup Language (HTML) and Document Object Model (DOM)by the W3C led to the triggering of a window of opportunity for coupling different solutions to the problem of web architecture development. This resulted in the establishment of WHATWG which appropriated the development of the Living Standard for HTML and DOM from the W3C. The loss of this development by the W3C led to an institutional change and a consequential focus on web architecture, the semantic web, the web of devices, and web and television (TV) standards. The second part of this chapter will discuss the development of web and TV standards and how the move to the W3C has served to rebalance interests in the global debate over copyright protection.

Global Standard-Setting in Internet Governance. Alison Harcourt, George Christou and Seamus Simpson, Oxford University Press (2020).
DOI: 10.1093/law-oso/9780198841524.001.0001

HTML and DOM Migrate to WHATWG

HTML is a mark-up language standardized by the W3C. It is used for creating web pages and applications. Tim Berners-Lee developed HTML in 1989 and released it as an open source protocol. It was later standardized within the IETF in 1993 by Berners-Lee and Dan Connolly. A competing HTML+ (Hypertext Markup Format) draft was designed at the same time by Dave Raggett. The IETF established an HTML working group in 1994 to work on HTML 2.0 in 1995. Berners-Lee was frustrated with the slow development of agreement within the IETF which was mainly due to competing interests. For this reason, he established W3C in 1994 at MIT to work on HTML development. The IETF HTML working group ceased to exist in 1996 and HTML 3.2 was published as a W3C Recommendation in January 1997, followed by HTML 4.0 in December that year. By 2000, HTML was confirmed as an international standard by the International Organization for Standardization (ISO).

A problem stream was created with the appearance of the extensible Markup Language (XML) which was developed within the W3C to replace HTML. In February 1998, the W3C released XML 1.0 as a W3C Recommendation. This was flanked in June of 1998 with a W3C workshop entitled Shaping the Future of HTML, led by Chairs Dave Raggett and Steven Pemberton.[1] The purpose of the workshop was to promote the transition to a new stage of HTML based upon XML tag-sets.[2] Work on the standard resulted in XHTML 1.0 which was published as a W3C Recommendation in 2000.[3] XHTML was a reformulation of HTML in XML syntax.

However, when XHTML 2.0 was introduced by the W3C in 2002[4] it was incompatible with the older HTML/WHTML versions. It also enabled the modularization of XHTML for different devices.[5] A number of browser companies argued that this incompatibility would have created sizeable upgrade costs for web developers across the Internet. For this reason, two browser companies, Opera and Mozilla, initially via a W3C mailing list, began working on alternative solutions for XHTML which were compatible with the older HTML4, XHTML1, and DOM2 HTML versions.

[1] http://www.w3.org/MarkUp/future/.

[2] For a comprehensive explanation for the difference between HTML and XML see: https://www.w3schools.com/xml/xml_whatis.asp.

[3] https://www.w3.org/TR/2000/REC-xhtml1-20000126/.

[4] http://www.w3.org/TR/2002/WD-xhtml2-20020805/.

[5] https://www.w3.org/TR/2001/REC-xhtml-modularization-20010410/.

In June 2004, Opera and Mozilla presented a position paper at a two-day W3C workshop.[6] Opera, Mozilla, and Microsoft were dominant in the browser software market at the time, with Apple recently entering the market with Safari (released in 2003). The key recommendation of the position paper was to extend HTML. It also recommended the use of JavaScript and XBL,[7] argued against device-specific profiles, and called for an open specification process with integrated features to be backed up with specific use cases. However, the proposal did not find favour with the majority of W3C members. A vote on the second day out of 51 attendees showed 8 in favour and 14 against. As explained by Sivonen (2007):

> the balance of power in the W3C had shifted from traditional desktop browser vendors to various other interest groups such as makers of software for mobile walled gardens and developers of 'rich client' technologies that could be deployed on intranets but that were not used by the general public on the Web. This had led to a situation where the focus was more on the 'Semantic Web', 'Web Services' and 'Mobile Web' than on what is usually considered 'the Web'. As a result, the development of the Web itself had been neglected.

Subsequently, a new organisation the WHATWG was established, led by Ian Hickson at Opera two days after the workshop in June 2004. A WHATWG website was created[8] and Opera, Microsoft, and Mozilla began to work externally to the W3C, with a public mailing list for technical discussion. The website stated that 'the working group will create technical specifications that are intended for implementation in mass-market Web browsers, in particular Safari, Mozilla, and Opera'. Microsoft was invited to participate in WHATWG but declined because WHATWG did not stipulate an intellectual property rights (IPR) policy.[9]

HTML 5.0 was adopted by IBM, Microsoft, and Apple. By 2004, HTML 5.0 saw adoption by Google, Apple, Opera, and Mozilla, which ran in parallel to WHTML 2.0. Ian Hickson was hired by Google in October 2005 to work permanently on the HTML 5.0 standard. Standardization of the Document Object Model (DOM) also migrated to the WHATWG at this time.[10] Mozilla

[6] http://www.w3.org/2004/04/webapps-cdf-ws/.

[7] XBL was only used by Mozilla. Mozilla long tried to standardize XBL within the W3C but no other browser adopted it. Mozilla abandoned XBL in 2012.

[8] https://whatwg.org/news/start.

[9] https://channel9.msdn.com/Shows/Microsoft+Conversations+with+J/Jon-Udell-Chris-Wilson-on-IE7-Ajax-and-web-standards.

[10] https://dom.spec.whatwg.org/.

and Opera were granted permission to publish HTML 5.0 as a W3C specification on a royalty-free basis in 2005. The most important feature of the HTML5 standard was that existing webpages would render correctly even when an older browser was used. XML, in contrast, would error when a fault was detected. Thus, a mistake on the part of a web developer would result in an unreadable webpage. However, the two standards, HTML and WHTML, continued to develop in parallel. HTML was adopted by Microsoft for Internet Explorer in 2008 and Microsoft refused to support XHTML from 2008 onwards.

Due to the large take-up of HTML5, the W3C decided to work on the standard after all. In a blog post entitled 'Reinventing HTML', Berners-Less stated:

> It is necessary to evolve HTML incrementally. The attempt to get the world to switch to XML . . . all at once didn't work. . . The plan is to charter a completely new HTML group. Unlike the previous one, this one will be chartered to do incremental improvements to HTML, as also in parallel xHTML. (Pilgrim, 2010:13)

Subsequently, in 2007,[11] a joint working group was established by the W3C to work with WHATWG on HTML 5.0 led by Ian Hickson. In January 2008, a working draft of the updated HTML5 specification was published by the W3C.[12] However, even though the W3C kept working on HTML5.2, it remained at Recommendation phase until 2014.

This represented a classic case of layering. Hacker argued that layering should lead to institutional change and eventually alter arrangements within the existing organization, which is evident, in this case, in the W3C. This is referred to in the institutionalist literature as policy drift which means that there will be a slow change in the original organization. Normally this occurs when the market changes and existing arrangements become less significant. The W3C, as a recognized standards organization, with formalized consensus-driven rules, and membership fees, ran a much slower agreement progress with a large degree of scrutiny by many members. By contrast, WHATWG's specification is fast-moving, developed between a small number of browser developers dominant in the market.

[11] http://www.w3.org/2007/03/HTML-WG-charter.html.

[12] https://www.w3.org/2004/04/webapps-cdf-ws/papers/opera.html and https://www.w3.org/TR/2008/WD-html5-20080122/.

In 2011, splits formed between the W3C and WHATWG again. From this time on, WHATWG continued to work on what it terms the 'Living Standard' for HTML and DOM. Initially the Living Standard was developed by Opera, Apple, and Mozilla, later joined by Google and Microsoft. By 2011, WHATWG changed the name of its HTML5 standard to HTML. The Living Standard is continually updated. The W3C continued to work on its version of the HTML5 standard in parallel, adopting some, but not all, updates from the Living Standard. The W3C outlines the differences between the specifications in an online document.[13]

In 2014, the W3C published HTML5 as a W3C Recommendation.[14] This was followed in 2015 with the publication of DOM4.[15] The specifications served as snapshots in time of the WHATWG Living Standard with W3C variations built in. Interviews with WHATWG members revealed frustration that the W3C simply 'cut, paste and altered' WHATWG specifications. Similar frustrations with fragmentation and lack of wider scrutiny into the development of the Living Standard were expressed by developers participating in the W3C.

In October 2015, the W3C established the Web Platform Working Group and dissolved the HTML working group. From 2016, feedback and comments on HTML drafts from developers were submitted to the Github repository.[16] This was followed by HTML 5.1 published as a W3C Recommendation in November 2016 and HTML 5.2 in December 2017. Microsoft formally joined WHATWG in 2017 when WHATWG revised its IPR policy.[17] Under its reorganization WHATWG integrated a royalty-free patent policy, a Review Draft stage, a code of conduct, the attribution license CC-BY, and continued collaboration with the W3C.

There have been repeated splits between browser companies over the use of layout engines. Most of these revolved around proprietary versus open source solutions. Web layout engines display web content from mark-up, style sheet, and user interface languages such as HTML, CSS, XUL, and JavaScript. Since 2013, Google and Opera use Blink, an engine used for the Google Chrome and Opera browsers.[18] Microsoft's Internet Explorer formerly used Trident and then EdgeHTML but announced a switch to Blink in December 2018. Mozilla's

[13] https://www.w3.org/wiki/HTML/W3C-WHATWG-Differences.

[14] https://www.w3.org/blog/news/archives/4167.

[15] https://www.w3.org/TR/2015/REC-dom-20151119/.

[16] https://github.com/w3c/html/issues. Github is also used for web platform tests: https://github.com/web-platform-tests/wpt.

[17] https://blog.whatwg.org/working-mode-changes.

[18] Blink is an open source Chromium project developed by mainly by Google, ASA, Adobe Systems, Intel, Opera, and Samsung.

Firefox uses the Gecko layout engine[19] and Apple's Safari uses WebKit. Trident, EdgeHTML, and Presto were proprietary browser engines whereas Blink and Gecko are open source. WebKit has been trademarked by Apple but is available under a BSD-form license. Microsoft switched to Blink in 2019 due to the large costs involved in the continued development of Edge.

Some in the industry claim that Google's Chrome is behaving increasingly like Microsoft's Internet Explorer in the early 2000s,[20] with an increasing number of Chrome-only sites available on the web.[21] When Chrome was released in 2008 it was dubbed as a compatible browser alternative to Internet Explorer which was, at that time, increasingly diverging from agreements in the W3C over HTML, particularly in the development of scalable vector graphics. However, Chrome is currently demonstrating increasing divergence.

There have also been accusations that Google deliberately attempted to disrupt Microsoft's Edge. In December 2018, a Microsoft intern, Joshua Bakita, accused Google of deliberately adding code to disrupt the compatibility of Edge on YouTube. As a former Microsoft intern, Bakita explained 'one of the reasons we [Microsoft] decided to end EdgeHTML was because Google kept making changes to its sites that broke other browsers, and we couldn't keep up'. He further claims that Google inserted a 'hidden empty div over YouTube videos' that disrupted Microsoft's hardware acceleration for videos.[22] Other browser operators show the same lag behind Google. For example, in 2018, Microsoft's update to the Web version of its Skype client only worked in Chrome and Edge.

Indeed, Google is dominant in the browser market with 60% of overall market share (Figure 5.1). Many browsers also utilize Google Chrome's Chromium, an open source web browser project, including Amazon Silk, Blisk, Brave, Opera, Vivaldi, and Samsung Internet. Microsoft Edge made the decision to use Chromium from 2019. This has raised concerns within the community that Google may be operating as a gatekeeper.

Arguably, the most important feature of HTML5 was that it integrated application programming interfaces (APIs). As Lee explains this allows for 'client-side persistent storage with both key/value and SQL database support; offline-application, editing, drag-and-drop, and network APIs, and cross-document messaging' (Lee, 2008:48). Notably, as there are two versions of

[19] Gecko is being developed by the Mozilla Project https://developer.mozilla.org/en-US/docs/Mozilla/Gecko.

[20] https://www.theverge.com/2018/1/4/16805216/google-chrome-only-sites-internet-explorer-6-web-standards.

[21] https://arstechnica.com/gadgets/2019/03/microsofts-new-skype-for-web-client-an-early-taste-of-the-browser-monoculture/.

[22] https://www.theverge.com/2018/12/19/18148736/google-youtube-microsoft-edge-intern-claims.

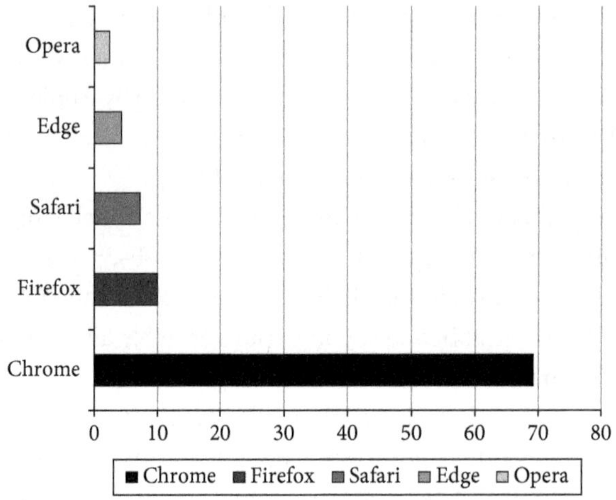

Figure 5.1 Web browser market share in 2019.
Source: http://gs.statcounter.com/browser-market-share/desktop/worldwide/#monthly-200901-201807

HTML5, the Living Standard and the W3C specification, this means that there are different APIs relating to each. Both the Living Standard and W3C have text track APIs, drag and drop APIs, web messaging and web storage which allows users to continue using a webpage when offline (e.g. for email or gaming), user interaction APIs, history APIs, and protocol handler registration. For the Living Standard, there are geolocation, web audio, web RTC, and web cryptography APIs. The W3C APIs[23] include cascading style sheets (CSS), media specifications (Encrypted Media Extensions (EMEs), media source extensions, scalable vector graphics, and web audio), graphics, fonts, networking, and security APIs.

The splits between and within WHATWG and the W3C led to concerted effort to forge a Memorandum of Understanding[24] between the two organizations. As W3C Chief Executive Jeff Jaffe said in a blog post, the memorandum agreed in May 2019 was 'motivated by the belief that having two distinct HTML and DOM specifications claiming to be normative is generally harmful for the community, and the mutual desire to bring the work back together'.[25] Under the memorandum, the W3C and WHATWG agreed to work together on HTML and DOM. It was agreed that WHATWG would maintain the HTML and

[23] https://www.w3.org/2018/12/webmediaapi.html.
[24] https://www.w3.org/2019/04/WHATWG-W3C-MOU.html.
[25] https://www.w3.org/blog/2019/05/w3c-and-whatwg-to-work-together-to-advance-the-open-web-platform/.

DOM Living Standards but would submit these as W3C Recommendations. This would 'facilitate . . . community work directly in the WHATWG repositories (bridging communities, developing use cases, filing issues, writing tests, mediating issue resolution)'.[26] The W3C would in turn cease to publish independent specifications based on HTML and DOM.

Standards and Content Protection

The mechanism of layering led to a situation where a new standard was introduced alongside an existing one. This came about due to stasis in existing HTML development within the W3C. This triggered a window of opportunity for coupling different solutions to the problem of web architecture development resulting in the Living Standard. The institutionalist literature on layering argues that this normally leads to institutional change. The W3C did indeed change as a result of its loss of HTML and DOM development to WHATWG with greater focus on web architecture, the semantic web, and the web of devices. One salient area of work is payments applications. Most work on the standardisation of online payment applications takes place in the Web Payments Working Group,[27] most importantly with development of the Payment Request API.[28]

An important new focus for the W3C is web and TV standards. The W3C started working on video on the web from 2007 and TV from 2008 onwards. Its TV Control Working Group, which met until 2017, worked on specifications that enabled web pages to browse and control TV and radio channels with audio and video.[29] Its Second Screen Working Group defines APIs to allow web applications to show web content on second screens or displays such as ebooks, mobile phones, laptops, and electronic billboards.[30]

APIs enable music and movie streaming on web browsers. This can also be done with plug-in add-ons like Microsoft's Silverlight and Adobe Flash, however HTML5 made it possible to standardize these features by building them into browsers. For this reason, technologies for copyright protection have also moved to standards bodies, most notably with development of the EME

[26] https://www.w3.org/blog/2019/05/w3c-and-whatwg-to-work-together-to-advance-the-open-web-platform/.
[27] https://www.w3.org/Payments/WG/.
[28] https://www.w3.org/TR/payment-request/.
[29] https://www.w3.org/TR/tvcontrol-api/.
[30] https://www.w3.org/2014/secondscreen/charter-2018.html.

at the W3C (W3C, 2016). Notably, the attractiveness of removing the need for a 'middle-man' (e.g. plug-ins, delivery set-top boxes) started to attract rights holders to the W3C such as the BBC, Cable Labs, Comcast, Netflix, and NBCUniversal, as well as associations such as the Motion Picture Association (MPAA), representing the major Hollywood film studios (and since 2019, Netflix) and the Recording Industry Association of America (RIAA), which represents corporations from the recording industry such as EMI, Sony, Universal, and Warner.

Content protection and standards have a long history and have developed in tandem (Khalifa and binti Yusof, 2012). Watermarking was developed for images in the late 1990s (Black, 2005:391). Digital watermarking was introduced from the mid-2000s, and was then expanded to audio and video. In 2006, the Digital Watermaking Alliance was established in the United States and Stirmark was developed as a standard for watermarking tests. Digital watermarking inserts bytes of data into 'noise' parts of content which is used for its identification. The watermark is undetectable to the user. Its primary use is for identifying and locating content, particularly within content libraries, but it is increasingly being used to block the uploading of content and for automatic take-down of content on user-generated content websites such as Facebook, YouTube, and Vimeo. The protection and blocking of content has always been controversial but has become particularly contentious with the passing of the EU's Directive on Copyright in the Digital Single Market in 2019. Article 17 of the directive (draft Article 13) mandates that online platforms use 'effective and proportionate' measures to prevent copyright infringement. This implies that technologies should be in use to identify and automatically take down content.

Similar to watermarking, fingerprinting uses an algorithm to examine and identify 'fingerprints' in files. These are checked against a database in a similar way that audio files are checked in music libraries. Napster was the first to use fingerprinting for identifying audio files in 2001. In 2006, audiovisual fingerprinting was developed. Large content libraries now use fingerprinting or watermarking to find all forms of content. Newspapers use fingerprinting to locate print content within their own archives. Nielson, the TV ratings agency, uses it to track audience viewing data. Fingerprinting is also used to manage user-generated content by social media platforms. When users upload memes or other videos to social media, fingerprints identify and label the content. Big Data companies use fingerprinting to collect information on social media user activity and provide them to companies for targeted advertising. Indeed, this technology is becoming so advanced that it is prompting highly innovative usage; for example, in 2015 Google filed a patent for fingerprinting signatures of cameras.[31] Facebook has since developed this technology to determine whether photographs uploaded by one user have been taken by another user.

Rights expression languages (REL) are used to read digital rights management (DRM) mark-ups within files (whether they be audio, audiovisual, documents, etc.). REL was standardized for interoperability from the mid-2000s. The eXtensible Access Control Markup Language (XACML) and the Security Assertion Markup Language (SAML) were patented by IBM and adopted by OASIS as access control languages in 2005.[32] They are used by Amazon, Microsoft, IBM, and Google. . The Telecommunication Standardization Sector, a division of the ITU (ITU-T), and the Organization for the Advancement of Structured Information Standards (OASIS) have an agreement over SAML.

Also adopted by OASIS, the eXtensible Rights Markup Language (XrML) was patented by ContentGuard[33] and is used by AT&T, Axiomatics, Balan, Microsoft, and Sun Systems as a basis for development of the MPEG-21 RELs and utilized by a range of users including Scribd.[34] In 2006, Automated Content Access Protocol (ACAP)[35] was standardized by the IETF as an REL developed by the World Association of Newspapers (WAN), the European Publishers Council (EPC), and the International Publishers Association (IPA) to prevent and regulate the way in which news aggregate sites reuse online newspaper stories.[36] The Associated Press chose a different REL, hNews, to tag its news content online.[37] ccREL, developed by the Creative Commons, determines rights for users which can include commercial use.

From 2006, Open Digital Rights Language (ODRL) was developed for reading XML as a royalty-free standard. ODRL was firstly used by the Open Mobile Alliance (OMA) for Mobile Communications for reading OMA DRMs. It was standardised by the W3C Community and Business Group in 2011. It is in use by a large number of outlets from ebooks to online newspapers. Work on standardisation continues in the W3C Permissions and Obligations Expression Working Group from 2014, one example being automating sets for financial data from stock exchanges. Another standard, PRISM Aggregator Message for Web Content (PAMW), was developed by Idealliance in 2014 for XML.[38]

[31] https://www.google.com/patents/US20150124107.

[32] https://www.oasis-open.org/committees/xacml/.

[33] See use by Google: https://www.google.com/patents/US9626668. For history on ContentGuard see Kia, 2010:291.

[34] https://www.scribd.com/doc/112242347/251/XrML?sh=02f31a347e603860.

[35] Automated Content Access Protocol.

[36] http://the-acap.org/FAQs.php.

[37] https://techcrunch.com/2009/07/29/aps-rights-management-drm-explained/.

[38] http://www.idealliance.org/specifications/prism-metadata-initiative.

This enables the slicing up and labelling of chunks of data as well as the protection of data for commercial use.

Historically, rights management technologies facilitated digital content protection and have been particularly important for copyright protection. The first forms of rights management began with audiovisual content protection for film and TV, namely CSS for DVDs and conditional access systems (CAS) for pay TV (Black, 2005:395). CAS consisted initially of the scrambling of cable, satellite, and terrestrial signals. Over time, rights management evolved into complex software packages with the advent of digital technology. Initially, DRM existed only to encrypted audio and audiovisual content but was later developed for documents, on-demand services, streaming, and gaming. DRM technology is proprietary, with companies mainly producing software for their own devices.

Providers wrap content with DRM when taking them from digital libraries and depositories (i.e. digital file servers) and providing them to platforms (e.g. a computer, iPod, Kindle, mobile phone, TV set, or other device). The software not only encrypts the content but also adds other information to it such as price, the identity of the user, and the type of rights permitted (e.g. viewing, printing, copying, transfer to other devices, time limits on use, etc.; Perzanowski and Schultz, 2017).

Each piece of content has a unique identifier. Standard-developing organizations (SDOs) were involved from early on in this process providing classifications for different forms of content. Audio content (mostly music) is identified by International Standard Musical Work Code (ISWC) and grid global release IDs; audiovisual content is categorized by International Standard Audiovisual Number (ISAN) and Unique Material Identifier (UMID) standards, and other types of content have a digital object identifier (DOI) or a persistent uniform resource locator(PURL)) such as academic journal articles. The registration agencies for music are the International Confederation of Societies of Authors and Composers (CISAC), the International Federation of the Phonographic Industry (IFPI), and ISO[39] and the Society of Motion Picture and Television Engineers (SMPTE) for audiovisual content. DRM for audio emerged in the early 2000s after the development of MPEG-4 in 1998 which enabled digital encryption for the first time.[40] Advanced Audio Coding (ACC) files,[41]

[39] ISO developed the International Standard Audiovisual Number (ISAN).

[40] The Moving Picture Experts Group (MPEG) is a self-regulatory professional forum put together in 1988 by the International Organization for Standardization (ISO) and International Electrotechnical Commission (IEC) to formulate audio and video compression and transmission standards. It operates under the ISO/IEC working groups. History: MPEG (1993), MPEG-2 (1995), MPEG-4 (1998), MPEG-A (2007) multimedia application formats MPEG-B (2006), and MPEG-V (2011).

[41] AAC is the audio standard agreed by the ISO and IEC under MPEG-4 specifications.

the successor of MP3, was used in devices such as Apple (iPad, iPhone, iPod, iTunes, and AppleTV), Android, DivX, DVB, HDTV, Nintendo DS, and Wii, Nokia phones, Sony's PlayStation and Walkman, and YouTube.[42] DRM was used to wrap ACC files from 2003 to 2009. During this period, there were four main DRM technologies in use for music files: Apple's FairPlay, Microsoft's PlaysForSure, and Realnetwork's Helix. A wider range of DRM technologies were developed for audiovisual (film and TV) programming.[43]

From 2003, DRM was developed for computer documents. It enabled the identification of the user, their role in the organization, and the computer they use to access documents. It also identified the form of documents (memo, internal communication, regulations, etc.). As with audio DRM, document software is proprietary. Examples are Microsoft's Rights Management Services, Oracle's Sealed Media, Documentum's Authentica, and Adobe's LiveCycle Rights Management. Lastly, DRM was developed for on-demand, online services as well as mass online multiplayer games (MMOGs). These include DRM technologies such as Adobe's Primetime, DivX, Google's Widevine (set-top boxes), Marlin[44] (PlayStation, Tivu, YouView), and the OMA's DRMs (for mobile phones).[45]

The market for audio content was dominated in the mid-2000s by Apple, Microsoft, and Realnetworks, which provided content from four main libraries, EMI, Sony, Universal, and Warner. As the DRM technologies affected consumers who downloaded music, activists were thus able to target their efforts towards a small group of corporations and specific laws. The key target was Apple's DRM, FairPlay. Designed by Veridisc in 2001, it was built into QuickTime for use by iTunes, Apple's App Store, Apple TV, iPad, iPhone, and iPod.

Apple was not the only company to run a propriety DRM system. In 2004, Microsoft created the DRM PlaysForSure[46] for Windows Media Player 10. However, unlike FairPlay, Microsoft made licensing available for other platforms in addition to PlaysForSure including Creative, Dell, HP, and Samsung.

[42] Other audio standards include TwinVQ, BSAC for general audio coding, HVXC and CELP for speech coding, MPEG-4 SLS, MPEG DST, and Audio Lossless Coding for audio compression, and for other synthesized uses (TTSI, SAOL, SASL, MIDI).

[43] Most CAS systems in Europe use digital video broadcasting (DVB), currently DVB Simulcrypt. Other CAS systems use Cisco, Conax, Eurocrypt, Irdeto, Nagravision, Viaccess, and VideoGuard.

[44] http://www.businesswire.com/news/home/20160908006794/en/Intertrust-Announces-Dolby-Vision-Compatibility-ExpressPlay-DRM.

[45] A wide number of DRM technologies are used by the OMA including Authenteca, Discretix, Irdeto, Philips, and Saffron.

[46] This was updated with PlayReady in 2007.

Realnetworks devised Helix DRM software for RealPlayer. From 2004, it started to provide Harmony software which allowed Apple iPod users to download music from RealNetworks' Rhapsody music store, converting Helix DRM for iPod use. In 2004, Apple accused Realnetworks of 'hacker tactics' when it introduced the Harmony software.[47] In 2007, Microsoft developed Coral as an interoperable DRM standard but Microsoft, RealNetworks, and Apple did not adopt it.[48] From 2011, Nokia introduced DRM to its MixRadio online service.[49]

As documented by many scholars, the problem with Apple's FairPlay in particular was that there was no interoperability, which locked customers into Apple devices. Most authors writing on this agree that the DRM technology enabled Apple's market dominance (Burkart and McCourt, 2006:13; Sharpe and Arewa Olufunmilayo, 2007). By the mid-2000s, Apple had gained 80% of the digital music market. This created a lot of discontent amongst other providers and consumers which led not only to high-profile activism but also to direct action in the form of large-scale piracy (Sharpe and Olufunmilayo, 2007:3). As Finnemore and Sikkink point out: 'deliberately inappropriate acts (such as organized civil disobedience) . . . can be powerful tools for norms entrepreneurs seeking to send a message and to frame an issue' (1998:897). This tactic was used in the form of the wide distribution by activist groups of software to decrypt audio files, such as Beale Screamer, DeDRMS, Hymn, NoteBurner, PlayFair, and QTFairUse.

Piracy imposed heavy costs for media players which needed to invest further in DRM technology to make hacking more difficult. This made the technology increasing expensive to develop. It produced a spiral effect in that, the more complex DRM became, the more difficult it was to implement by platforms. The unbundling of the software also became increasingly complicated for users. As Steve Jobs stated: 'any company trying to protect content using a DRM must frequently update it with new and harder to discover secrets. It is a cat-and-mouse game.'[50]

[47] http://www.macworld.com/article/1035244/applereal.html.

[48] http://www.infoworld.com/article/2681984/applications/vendors-look-to-tame-drm-standards-with-coral.html.

[49] MixRadio was expanded to use for Android and iOS after sale to Line Corporation in 2015 but closed in 2016.

[50] http://www.apple.com/kr/hotnews/thoughtsonmusic/.

Wide Public Profile of the Issue of DRM

Most debate over copyright protection took place within national contexts. In 2004, the French group Virgin Mega (subsidiary of the Lagardère) filed a complaint with the French Competition Council about Apple's refusal to license FairPlay to other providers.[51] Customers downloading music from Virgin Mega couldn't play music on iPods. Apple's refusal to grant access was seen as an abuse of dominant position. In February 2005, the French consumer group, Union Fédérale des Consommateurs-Que Choisir, filed a lawsuit against Apple in France. This leveraged France to enact a 2006 law to enforce interoperability of DRM technologies (Rimmer, 2007:83–5).[52] By June 2006, the Consumer Ombudsmen offices of Denmark, Norway, and Sweden formally accused Apple of transgressing consumer contract law. In the United Kingdom, the record industry trade body, the British Phonographic Industry (BPI), responded to pressure from the Open Rights Group and the British Software Alliance (BSA) to request Parliament to mandate interoperability.[53] The German consumer group, Ferbraucherzentralen, and Finland's Kuluttajavirasto requested interoperability by the end of 2006.

Meanwhile in the United States, campaigning by digital rights groups led to a class action brought against Apple on behalf of over 8 million consumers and hundreds of retailers.[54] The claim was that the lack of interoperability was anti-competitive as it shut off market access. This combined pressure led to the publication of an open letter by Steve Jobs on the Apple website in February 2007 entitled 'Thoughts on Music'.[55] It called on the 'Big Four' music companies to sell their music without DRM. In April 2007, EMI backed down and permitted download of its music without DRM. By 2009, all downloadable music from the Big Four was downloadable without DRM technology. Amazon started to provide downloadable DRM-free music.[56]

Negative media comment around DRM was utilised by digital rights groups to garner support against DRM. A prime example of this was in 1999 when the recording industry published an open call to hackers and other users to crack the digital watermark technology, SDMI, which was in wide use at the time. A $10,000 reward was offered. The watermark was broken by Professor

[51] http://www.autoritedelaconcurrence.fr/pdf/avis/04d54.pdf.
[52] LOI n° 2006-961 du 1er août 2006 relative au droit d'auteur et aux droits voisins dans la société de l'information.
[53] http://www.ft.com/cms/s/0/cd0034d4-fb42-11da-b4d0-0000779e2340.html.
[54] https://casetext.com/case/in-re-apple-ipod-itunes-antitrust-litig-5.
[55] http://www.apple.com/kr/hotnews/thoughtsonmusic/.
[56] However, at a later date, the users' ID was embedded into the download to track use.

Ed Felten from Princeton University. Rather than collecting the reward, he planned to publish a paper on the weakness of the technology. The Recording Industry Association of America (RIAA) and the Verance Corporation subsequently threatened him with legal action claiming transgression of the Digital Millennium Copyright Act (DMCA). Felten filed a counter-case for permission to publish the paper in the interests of academic freedom under the DMCA's exemption for encryption research (section 1201(g)). A high-profile online support campaign on behalf of Felten was led by the Electronic Frontier Foundation supported by dozens of digital rights groups. They were highly organized and focused on a single target, the RIAA. Under the pressure, the RIAA and Verance withdrew the case.

This battle over DRM between rights holders and activists continued within the US when in 2000, eight film studios[57] challenged three men over the distribution of a software program called DeCSS which was used to decrypt the CSS DRM software and pirate films.[58] The ACLU publicly supported the defendants in an appeal on the grounds of free speech. The defendants claimed that CSS was so easy to crack that it should not qualify for contravention of the 1998 DMCA. Although the case and a subsequent 2004 CSS case[59] were lost, the 'easy to crack' statement was so widely publicized by digital rights groups and in the media that it permanently damaged the reputation of the software which is no longer in use. The Electronic Frontier Foundation issued a Pioneer Award to Dan Gillmor, Beth Givens, and Jon Johansen public accrediting them as the authors of DeCSS. Gillmor and Givens were working for the ACLU and Privacy Rights Clearinghouse respectively at the time.

A similar public campaign was mounted against audio DRM. Most ebook lending services use the Overdrive library service in Cleveland, Ohio. Adobe's Content Server DRM software encrypts access for Android, Amazon Kindle, Windows Phone, and iOS devices (e.g. the iPad). Dmitry Sklyarov developed an advanced e-book processor for the Russian company ElcomSoft which decrypted Adobe's DRM, allowing users to make copies of their downloaded purchases. Sklyarov was taken into custody in the United States after returning from a DEFCON conference in 2001. Charges against him were pursued by the federal prosecutor, Robert Mueller, with the support of the Association of American Publishers (AAP). Sklyarov's detention provoked a large, public

[57] Universal Studios, Paramount Pictures, Metro-Goldwyn-Mayer Studios, Tristar Pictures, Columbia Pictures, Time Warner Entertainment, Disney Enterprises, and 20th Century Fox.

[58] http://partners.nytimes.com/library/tech/00/01/biztech/articles/15disc.html.

[59] In the 2004 *DVD Copy Control Association v Bunner*, programmers and webhosts were successfully accused of distributing pirating software for DVDs.

campaign including street protests, a central website (freesklyarov.org), and mailing lists for organized protests as well as boycotts of Adobe software (Postigo, 2012:236).[60] Sklyarov was defended by Robin Gross, an attorney with the Electronic Frontier Foundation (EFF), and eventually released on bail and returned to Moscow. A federal court found that ElcomSoft may have circumvented the software but Sklyarov was not guilty as charged. The outcome differed from the preceding CSS cases, which indicted individuals, signalling that naming and shaming tactics may have swayed court opinion. This is yet to be tested at the level of the Supreme Court where the EFF recently aspires to challenge the 'anti-circumvention' and 'anti-trafficking' provisions of the DMCA.[61]

DRM and the W3C

The conflict over DRM shifted away from the national level to SDOs with the introduction of HTML5. This began in 2012 when the W3C announced it was developing technical standards for EME in advanced HTML5. The proposal was formally proposed by David Dorwin from Google, Adrian Bateman from Microsoft, and Mark Watson from Netflix (W3C, 2013; Dorwin et al., 2015). Interviews showed that although browsers and rights holders were involved from early on in decision-making, with a 2010 W3C workshop on protected content and a 2011 use case, digital rights groups did not become aware of W3C activities until EME was formally proposed. The standard was for an API for streaming music and movies on web browsers precluding the need for plug-in add-ons like Microsoft's Silverlight[62]. The aim is to encrypt audio-video content, such as film and TV programming, for download. The removal of the need for a 'middle-man' (e.g. plug-ins, delivery set-top boxes which would need to decode DRM, etc.) made EME highly attractive to rights holders, who would then need only deal with browser companies. It offered a finer degree of control and made DRM more affordable.

This attracted a number of large rights holder corporations such as Comcast, Cable Labs, and the BBC to the W3C as well as the MPAA and the RIAA, which joined the W3C in 2014. In interviews, browser companies claimed that the push for the API came from rights holders initially as they needed to integrate DRM into streaming in any case. But many browser companies also produce

[60] Postigo noted wide support amongst usenet newsgroups such as 'comp.text.pdf' as well as sites frequented by the hacker community such as slashdot.org and cryptome.org.
[61] https://www.eff.org/document/1201-complaint.
[62] https://www.w3.org/2016/03/EME-factsheet.

DRM technologies such as Adobe, Apple, Google, and Microsoft signalling clear incentive. Mozilla formally opposed EME but has adopted it due to fear of competition from other browsers. It is integrating an Adobe DRM technology for content downloads but, differently to Chrome and Internet Explorer, one which will not have access to users' hard drives or networks.[63]

Digital rights activists began organizing to oppose the standard.[64] The main tactic was to link opposition to EME to previous anti-DRM campaigns by labelling EME as 'DRM4 for HTML5' and to draw the W3C into wider national debates over DRM. In this respect, EME was the subject of specialist blogs and fora and negative reporting in the technology media. Street protests were organized outside the W3C office.[65] Support was leveraged from the United Nations Educational, Scientific and Cultural Organization (UNESCO)[66] and the European Parliament.[67] However, the issue drew little attention from the wider public and protests have drawn few people, particularly in comparison to large national protests organized, for example, against the Anti-Counterfeiting Trade Agreement (ACTA), the Stop Online Piracy Act (SOPA), and the proposed Preventing Real Online Threats to Economic Creativity and Theft of Intellectual Property Act (PIPA).

The digital rights group EFF joined the W3C as an official member in order to participate in decision-making in 2014. However, the barriers for entry to SDOs for digital rights groups are high in terms of cost and the technical expertise required. As Cory Doctorate explains, the forum shift to the W3C produced a closed door effect as

> the W3C works in a uniquely transparent way. Specifications are largely developed in public and most groups have public minutes and mailings lists. However, Member confidentiality is a very valuable part of the W3C process.

[63] Firefox is using Adobe's Content Decryption Module (CDM): https://hacks.mozilla.org/2014/05/reconciling-mozillas-mission-and-w3c-eme/. As an article in PC World commented: 'Instead of baking in the CDM solution like other browsers, Mozilla is also giving users a high degree of control over the CDM just like any other plugin.' http://www.pcworld.com/article/2922116/firefox-38-arrives-with-contentious-closed-source-drm-integrated-by-default.html.

[64] Interviews with activist organizations showed that they became aware of the issue due to alerts from W3C members. W3C member Harry Halpin stated on Twitter that he would resign from W3C if it recommends EME, adding 'DRM is not part of the web we want' and he subsequently left in 2017. Active W3C members Ian Hickson and Boris Zbarsky publicly opposed the standard: https://lists.w3.org/Archives/Public/public-html/2012Feb/0274.html; https://lists.w3.org/Archives/Public/public-html/2012Feb/0352.html.

[65] https://www.eff.org/deeplinks/2016/03/scenes-anti-drm-protest-outside-w3c; http://dustycloud.org/blog/memories-of-a-march-against-drm/.

[66] https://assets.documentcloud.org/documents/3535173/Eme-Letter-Frank-La-Rue.pdf.

[67] The European Parliament discussed EME within a panel entitled 'Technology making law' on 15 October 2013 in Brussels. https://ec.europa.eu/digital-single-market/en/news/technology-making-law.

> That business and technical discussions can happen in confidence between members is invaluable to foster broader discussion, trust and the opportunity to be frank. The proceedings of the HTML Media Extensions work are public however, discussions amongst Advisory Committee members are confidential.[68]

The main civil society input took place in the United States by the EFF, the Free Software Foundation (FSF), the Centre for Democracy and Technology (CDT), and the Open Source Initiative (OSI). There were, however, differences in approach between the groups and no unified position. The EME campaign run by the EFF received support from the European Digital Rights initiative (EDRi),[69] the Open Rights Group,[70] and La Quadrature du net[71] in Europe.[72] Most groups agreed, however, on one central argument which is that the purpose of HTML is to provide a base language for web developers in order to maintain the openness of the web.[73] EME is viewed as incompatible with interoperability[74] and innovation.[75] Civil society groups point out that EME stems development of further browser software. As Doctorow stated: '[i]f other W3C standards worked like EME, there would have been no Mozilla' (2016). The EFF specifically addressed the issue that circumvention efforts may be subject to prosecution under the DMCA. The EME standard stipulates which browsers can carry content products provided by rights holders and which cannot; only browsers which adopt the standard can stream content.

A letter was sent to Tim Berners-Lee on 24 April 2013 coordinated by the FSF's 'Defective by Design' campaign,[76] signed by a large number of digital rights and open source groups.[77] The W3C's official response was that EME

[68] https://www.w3.org/blog/2016/06/perspectives-on-security-research-consensus-and-w3c-process/.

[69] https://edri.org/copyfail-9/.

[70] https://openrightsgroupbirmingham.wordpress.com/2017/04/10/email-w3c-stop-drm-becoming-web-standard/.

[71] https://www.laquadrature.net/en/node/8255.

[72] https://www.eff.org/deeplinks/2016/11/world-wide-web-consortium-crossroads-arms-dealers-or-standards-setters.

[73] https://www.w3.org/TR/html5-diff/.

[74] https://www.eff.org/deeplinks/2016/01/you-cant-destroy-village-save-it-w3c-vs-drm-round-two.

[75] https://www.eff.org/deeplinks/2016/03/interoperability-and-w3c-defending-future-present.

[76] http://www.defectivebydesign.org/sign-on-against-drm-in-html.

[77] April, Bristol Wireless, Call The Ninja, the Canadian Pirate Party, Cellularity Creative Commons, Electronic Frontier Foundation, Fight for the Future, Free Culture Foundation, FSF, Free Software Foundation Europe, Free Software Foundation India, Free Software Foundation Latin America, Gatsby Media, IT for Change, Junge Piraten, JustOutSource.it, Modern Poland, Oeko.net, Open Knowledge Foundation, Open Technology Institute (of the New America Foundation), Perpinux, Pirate Parties International, TheWayoftheWeb Ltd., Torservers.net, Pirate Party of Sweden, and the UK Pirate Party. See: http://www.defectivebydesign.org/sign-on-against-drm-in-html.

is not a DRM as it does not 'create nor impose a content protection or Digital Rights Management system. Rather, it defines a common API that may be used to discover, select and interact with such systems as well as with simpler content encryption systems.'[78] The W3C's view is that the standard is neutral and is not designed to support a specific DRM standard but can support many different ones. It could be utilized by either the major web browsers, such as with Google's Widevine DRM in Chrome or Microsoft's PlayReady in Internet Explorer, and DRMs designed by large content providers. The only requirement is that browsers implement Clear Key which enables encryption of media content. Browsers not supporting the key will not be able to unlock encrypted media. This response further diffused lobbying efforts, creating institutional ambiguity for activism.

Groups came out in support of the EFF in its protest against EME[79]. These included the blind people's rights organizations and higher educational institutes. The Royal National Institute for Blind People (in the United Kingdom), Media Access Australia, Vision Australia, Benetech, and SSB Bart (in the United States) argue that EME restricts development tools that can adapt video for disabled use (such as colour blindness or descriptive audio). Higher education and research institutes, including the Deutsche Nationalbibliothek, the Eindhoven University of Technology, Kings College London, Lawrence Berkeley Lab, Open University, Oxford University, and blockchain companies such as Ripple, Ethereum, and Blockstream, object to EME as it prevents research into defects in browser software. The EFF campaign was also supported by open source companies and associations such as Hypothes.is, Dublin Core, and Vivliostyle along with independent researchers,[80] key members of WHATWG,[81] and sector online news sites such as Privacy News Online,[82] Techdirt,[83] and the Software Development Times.[84]

The argument from open source groups was that EME exposes consumers to high risk as DRM is downloaded to their personal computers. Open source software is meant to be subject to scrutiny. EME is open to scrutiny under W3C

[78] https://www.w3.org/2016/03/EME-factsheet.html.

[79] https://www.eff.org/deeplinks/2016/11/world-wide-web-consortium-crossroads-arms-dealers-or-standards-setters.

[80] https://www.eff.org/deeplinks/2016/03/security-researchers-tell-w3c-protect-researchers-who-investigate-browsers.

[81] https://blog.whatwg.org/drm-and-web-security.

[82] https://www.privateinternetaccess.com/blog/2017/07/encrypted-media-extensions-copyright-drm-end-open-web/.

[83] https://www.techdirt.com/articles/20170403/23575537077/unesco-says-adding-drm-to-html-is-very-bad-idea.shtml.

[84] http://sdtimes.com/drm-concerns-arise-as-w3cs-tim-berners-lee-approves-the-eme-specification/.

development but the DRM it enables is not. As an FSF interviewee stated: 'DRM contracts a sphere of influence over that part of your computer which makes DRM use, entrench and legitimise the use of your computer. It muddies the waters.' Camri observed that this enabled

> user-specific information being disclosed to the EME (and thereby also the streaming media provider); and (2) . . . enable a third party to track the end user's viewing habits and/or their movements on the internet. . . This raised the possibility that companies providing access to copyrighted content via web streaming over HTML5 could potentially gather users' personal information and track them or target them with advertising. (2016:9)

As explained by the free software developer Ruben Rodriguez:

> If the DRM components of an otherwise freely licensed browser are distributed under a proprietary licensing scheme, then the browser is no longer actually free software, and cannot be distributed by projects that only distribute such. If to circumvent that—as Mozilla Firefox currently does—the non-free components are downloaded separately by the browser itself, then we have a backdoor that—even if we ignore the important security implications—robs the user of their control of their computation. It also exposes them to legal risk, as they are now in possession of DRM encumbered software they didn't request and are forbidden to audit. The worst case is when the DRM components are licensed under a free software license—as would be possible with ClearKey or other implementations—and thus included in the browser itself. The distribution of said browser under a free license would give users a false impression that it was safe to study, modify and redistribute the code. In reality, doing so would expose them to DMCA threats.[85]

This quotation demonstrates how difficult the issue is to communicate to wider audiences, taking into account the technical expertise needed to understand agreements. Notably, there was no support for the EME campaign from business associations such as Business Europe and MARQUES, wider organizations such as Doctors Without Borders, or even the BSA, who campaigned against ACTA. A survey and interviews with groups which had supported ACTA showed that few in Europe had even heard of EME, including Digitale Gesellschaft and the Panoptykon Foundation. Only after EME had

[85] https://github.com/w3c/encrypted-media/issues/305.

been formally adopted by the W3C did the Just Net Coalition object to the standard.[86]

Technical arguments provided by the FSF and EFF were further hampered by divisions between experts internal and external to the W3C and in technology publications.[87] This, coupled with lack of accessibility to decision-making, made the mobilization of public support almost impossible. As one open source lobbyist interviewed said:

> It may have been an organizational failure on our part but a lot of people see this as a less compelling argument and less important than legal acts such as the DCMA. DRM was once a high-profile struggle. But EME has been the hardest to message project I have ever done.

Clearly, the move to the W3C rebalanced actors participating in the global DRM debate. Activists were unsuccessful in preventing work on EME and it was formally adopted by the W3C in July 2017. The EFF called upon W3C members to adopt a 'covenant' to prevent lawsuits against those developing software which would bypass digital locks and/or publishing and distributing proprietary codes without permission from the original developers.[88] It reported support from 'a global, diverse coalition of commercial firms, nonprofits and educational institutions', including Stanford University, the Open University, University of Southampton, Pontifical Catholic University of Rio de Janeiro, the payment company Ripple, the Hypothes.is project, and 200 security researchers.[89] The EFF also called upon software developers, the FSF, and OSI for support.[90] However, a covenant was never adopted. Instead, the W3C released Guidelines for Vulnerability Disclosure Programs to protect security and privacy researchers on 27 January 2017.[91] Although the guidelines have received some attention in the specialist press,[92] other DRM areas such as fingerprinting, watermarking, and REL have not been addressed by digital rights groups. W3C members stated in interviews that they 'were

[86] https://lists.w3.org/Archives/Public/public-html-media/2017Apr/att-0041/W3C_EME_objection_endorsements.pdf.

[87] See, for example, a March 2017 Arstechnica article which supports EME: https://arstechnica.co.uk/business/2017/03/drm-in-html5-is-a-victory-for-the-open-web-not-a-defeat/.

[88] https://www.eff.org/deeplinks/2016/03/interoperability-and-w3c-defending-future-present.

[89] https://www.eff.org/deeplinks/2016/03/security-researchers-tell-w3c-protect-researchers-who-investigate-browsers.

[90] https://www.eff.org/deeplinks/2016/03/security-researchers-tell-w3c-protect-researchers-who-investigate-browsers.

[91] https://www.w3.org/2017/01/GVDP-factsheet.html.

[92] http://www.linux-magazine.com/index.php/layout/set/print/Issues/2013/157/Web-Cryptography-API/(tagID)/50.

warned by the W3C that the EME issue might 'spill over' to (other working groups) but there was nothing'.

Conclusion

This chapter has offered an understanding of layering with a case study on browser development. Standards for browser development have not followed an incremental or consistent institutional path. Instead, two institutional developments occurred. The first part of the chapter explained how since its establishment in 2004, WHATWG assumed the development of HTML and DOM with the Living Standard. This is consistent with work on layering and policy drift which sees a push for reform due to changes in the market and a need for innovation providing a window for change.

The loss of these standards provoked a move by the W3C to development on web architecture, the semantic web, the web of devices, and web and TV standards. The second part of the chapter showed how spillover to SDO fora results in a redistribution of power, one that empowers private sector governance. This occurs when formerly highly politicized issues are reframed in technical terms and decision-making is removed from the level of the state or bilateral treaty agreement to technical fora where representation is low. As Princen notes, just because actors are able to place issues onto agendas does not mean that they will succeed (2007:28). As shown by the DRM case, national legal environments are a powerful tool in moulding the shape that change can take. The DRM case is just one sign of increasing spillover of policy domains to SDO fora which have historically belonged to the state.

6

802.11ax

Technical Standards-making, the Unlicensed Spectrum, and the Future of WiFi

Introduction

Wireless communication is a central component of twenty-first century daily life, providing users with access to a gamut of audiovisual services from mundane voice telephony to social media and the streaming of broadcast and broadcast-like services, consumed live as well as later. It is likely that in the future the wireless environment will also play a vital part of the machine-to-machine communication of the Internet of Things (IoT). Thus, the airwave spectrum across which these services are provided is a highly valuable resource, not least because of its relatively scarce nature. Wireless communication services are made available through two main means: licensed operator provision to customers and, in contrast, unlicensed WiFi access. The openly accessible WiFi space—for the most part freely available to users—has become a vital way of accessing the Internet through mobile communications devices, laptops, and tablets in homes, businesses, and indoor and outdoor public spaces. It has also become a zone of increasing interest to licensed mobile broadband providers as demand for wireless services—and thus infrastructural capacity—has burgeoned. The unlicensed spectrum has become increasingly crowded and securing effective, high-quality communication within the 2.4 and 5GHz bands has become a core problem. This chapter utilizes the multiple streams (MS) framework in combination with insights from the literature on policy mechanisms (Falleti and Lynch, 2009) to explore the efficacy of the IEEE 802.11ax technical standard for WiFi as a policy process aimed at providing a solution that could be coupled with this problem stream.

The unlicensed spectrum is interesting in that, bar the assignment of the 2.4 and 5GHz bands by communications regulatory authorities, initially in the United States, its development and functioning has been underpinned by private technical standards-making conducted for the most part in the Institute of Electrical and Electronics Engineers (IEEE). Here, the 802.11

Global Standard-Setting in Internet Governance. Alison Harcourt, George Christou and Seamus Simpson, Oxford University Press (2020).
DOI: 10.1093/law-oso/9780198841524.001.0001

family of standards, developed for WiFi services, has historically dominated the functioning of the unlicensed bands in question. The chapter focuses on the development of the latest in this family—802.11ax—which was created in response to the need to deliver faster, more efficient WiFi communication in increasingly densely occupied unlicensed spaces. The chapter shows the process by which 802.11ax, as a constituent of the policy stream, was coupled with the unlicensed spectrum problem stream. The case is important since it shows how micro-level mechanisms (Falleti and Lynch, 2009) can function within international private standards-developing organizations (SDOs), thereby affecting the course of technical standards-making. Specifically, the existence of a highly controversial Special Interest Group (SIG) related to the development of 802.11ax impeded, albeit temporarily, the smooth development of the standard. The DensiFi SIG went beyond the accepted rules of the IEEE in that it was judged to have created so-called dominance in the development of 802.11ax.

Rather differently, but also in line with the understanding of Falletti and Lynch (2009), the chapter also shows how factors exogenous to the IEEE, related directly to the future unlicensed bands in question, impacted upon the 802.11ax process. Here, a parallel process of technical standards-making in the alternative international private SDO setting of the Third Generation Partnership Project (3GPP) served to heighten the need to speed the development of 802.11ax as a policy solution to the unlicensed spectrum problem. The specific context of this is that 802.11ax needed to be completed in order for it to become an alternative to the LTE-LAA standard developed by licensed mobile broadband operators in 3GPP wishing to extend their presence in the 5GHz unlicensed band. At the time of writing, 5G needs to harmonise 23 standards but has only thus far managed 2 agreements. Achieving completion would allow 802.11ax to be incorporated into a revised EU standard for local area wireless communication. The chapter thus shows how the unlicensed spectrum political stream comprised the LTE interest group and the EU politico-legislative context. A third element of the political stream was the strong consumer public preference (or consumer public mood, as understood by Zahariadis, 2005) to be able to use the latest WiFi services through their electronic devices. The chapter shows how the combining of the three streams led, in these often challenging circumstances, to development of 802.11ax.

The chapter is structured as follows: the next section describes briefly the emergence, character, and growth in importance of the 2.4 and 5Ghz unlicensed spectrum bands. In particular, it shows the significance of the IEEE's role in standards development for what has become known as WiFi. The following section provides an account of the development of the 802.11ax standard internally. It pays particular attention to the dominance controversy as a micro mechanism in the development of 802.11ax, in the process illustrating how the high stakes world of Internet standards development can be affected by systemic shocks. Following this, attention turns to the extra-IEEE context and the political stream of the 802.11ax process. Analysis is undertaken of the pressures which the development of 802.11ax faced in respect of LTE standards-making interests in the form of licensed mobile providers aiming to exploit their entry into, and use of the unlicensed spectrum. It also illustrates the importance of supranational—that is EU—legislation making and, rather differently, consumer public preferences in the political stream that led to the coupling of 802.11ax as a policy stream response to the unlicensed spectrum problem stream.

More broadly, the case illustrates the importance of private technical standards-making in wireless communication in three key respects: first, the regulation and maintenance of interference minimal, good order communications in the unlicensed spectrum; second, the commercial imperatives of ensuring that new standards-making develops expeditiously in line with—and in the service of—new electronic devices, enabling advanced wireless communication; and third, the role that micro-level mechanisms - (in this case adversarial) interest group pressure, legislative steer, and consumer/market pressures - can play in shaping the first two aspects.

The Problem Stream: The Unlicensed Spectrum, and Innovation in WiFi Standards to Deliver Better Services

The unlicensed spectrum environment, pioneered in the United States, was originally very different from the one recognized today. The Federal Communications Commission designated spectrum from three parts of the industrial, scientific, and medical (ISM) bands (in 902–928MHz, 2.4–2.483 GHz, and 5.7–5.85GHz), rather unflatteringly referred to in the media as 'garbage bands' (Economist, 2004) for highly specific purposes related to industrial process emissions and military uses, as well as the operation of some domestic appliances, notably cordless telephones and microwave ovens. However, in 1985, the use of these bands was loosened considerably, such that users were not required to have licenses to operate within them.

With the release of a further 300 MHz of spectrum in the 5GHz band by the Federal Communications Commission (FCC) and rapidly expanding demand to access the Internet, the context for the growth of WiFi was set in the United States, which continues to lead in terms of the number of established WiFi networks (World Economic Forum, 2016). Other countries soon followed suit. In conjunction with fixed broadband networks, WiFi has become a technology that provides users across many parts of the world with high-quality access to the Internet (Fogg, 2018). Global WiFi-enabled equipment shipments rose from 1.58 billion to 4.91 billion between 2012 and 2017 (Statista, 2019). By 2017, so widespread had the deployment of WiFi become that access to it was being discussed as a human right in commercial and political circles (see Powar, 2017). However, it is the case that the number of WiFi networks varies and is related to Internet access. Here, the African continent lags behind North America and Europe.

Unlicensed spectrum environments raise important considerations about their governability and governance. Open entry spaces, if they are to be effective, require self-regulatory usage protocols to be developed among those inhabiting the space the spectra provide. This involves adherence to highly practical conventions to set out how communications should occur. As these become established, it is inevitable that the practicalities of unlicensed communications become couched in cultural and normative understandings around key matters, such as expected behaviour in the spectrum, and the nature and quality of services and user experiences, and expectations held about it. In the local wireless area network environment in which WiFi developed, technical standards have played a vital role in setting the rules of communication as well as the cultural values associated with it.

The unlicensed wireless local area network (LAN) standards space is interesting in a number of respects. It provides evidence of historical and relatively stable incumbency in private international standards-making. Here, the IEEE's 802.11 wireless local area network Working Group (WG) developed successive standards which set the parameters for equipment seeking to operate in the space. The IEEE 802.11b standard was introduced in late 1999 and the IEEE 802.11a standard was ratified in early 2000 (Greenstein, 2007:14). The 802.11a standard utilized frequencies in the 5 GHz band, while 802.11b was designed to operate within the 2.4 GHz spectrum band (authors' interview, 2016). In August 1999, the Wireless Ethernet Compatibility Alliance (now much expanded into the WiFi Alliance)—comprising Intersil, 3Com, Nokia, Aironet, Symbol, and Lucent—agreed to begin certification of products which complied with the 802.11b standard. In order to give the technology a user-friendly name, WiFi was chosen (later sometimes retrofitted with the full name 'wireless fidelity') because of its visual and phonetic closeness to the popular 'HiFi' term of the time (CableFree, 2017). This and the subsequently developed versions of the 802.11 standard thus became known as WiFi. A key moment in the development of WiFi occurred when Apple agreed to incorporate a Lucent-made WiFi slot (called AirPort) into its iBook computers in 1999 (Economist, 2004).

WiFi was not the only system operating in unlicensed wireless local area network (WLAN) bands, yet its underpinning technical standards grew to dominate as Internet service provision and access surged from the late 1990s. Culturally, WiFi can be seen as having been pivotal to the development of the local WLAN with simultaneously a free market and public communications commons character. Entry was free for equipment vendors and service providers of every size. For users, WiFi was widely seen as a free Internet access service, yet like a lot of online activity, WiFi was in fact an attractive business opportunity. Being WiFi compatible came to be viewed as not just a selling point but later a prerequisite of quality, of staple Internet access for all consumer devices such as laptops, tablets, and mobile phones. Equally, the manufacture and sale of technical equipment related to signal transmission—notably wireless routers—became a major business opportunity.

The technical standards-making environment of the IEEE provided both a context for, and a reflection of, these developments. As noted in Chapter 3, the IEEE is an environment in which the pursuit of a common understanding of technical excellence sits alongside the imperative to reach agreement on a common technical platform from which the development of state-of-the-art electronic equipment and service provision can proceed on a commercial basis. The need to create as open an operational environment as possible through WiFi was reflected in the functionality of the 802.11 family of standards and made a core, underpinning contribution to the operational and cultural character of WiFi in the WLAN environment. This is illustrated by the technical feature of contention-based access to channels through the carrier sense multiple access with collision avoidance technical protocol (CSMA/CA) (Jeon, Niu, Li, Papathanassiou, and Wu, 2015). In this process, equipment using the WiFi bands undertake an automatic clear channel assessment in deciding whether or not to transmit a signal (Kwon, Jeon, Bhorkar, Ye, Harada, 2016). A key element of contention-based communication is the use of listen before talk (LBT). This is an algorithm which functions according to a listening time duration to detect the presence of other users, whereby if none is detected, transmission of the signal duly proceeds. The waiting process is termed random back-off. It is bolstered by the so-called exponential back-off mechanism which responds to a situation of signal collision by automatically increasing the random back-off interval before sending its signal (Dynamic Spectrum Alliance, 2015).

This technical system for ordering WiFi communication, in layperson's terms, can be seen as characterized by 'communicative politeness'. It created an open environment which attempted to use technical parameters to address the potential problem of signal interference. This amounted to a less than perfect compromise which delivered the openness of WiFi, the quid pro quo for which might occasionally be reduction of the quality of signal—and thus service—provided to the user. However, WiFi developed a communications culture which tolerated this, justified in significant part by access being free to users.

Nevertheless, a significant problem around quality of service emerged as WiFi became increasingly popular. In response, efforts were made in the IEEE via technical standards-making to enhance the speed and reliability of WiFi. The need to do this was dramatically heightened as users demanded both WiFi and licensed mobile broadband Internet access due to the development of the 'smart phone'. The growing interest in WiFi was part of a broader spectrum policy problem around the search for increased spectrum capacity to deliver fixed wireless and mobile advanced communications services.

A key issue for WiFi has been its relationship with licensed mobile wireless communications. In one respect, the two have been seen as competitors and their relative performances have been compared, in terms of Internet upload and download speeds. In a 2018 survey, for example, Fogg found that in countries with so-called 4G licensed mobile services, 63% of users experienced faster download speeds on licensed mobile services than WiFi (Fogg, 2018:11).

Yet, licensed mobile and unlicensed local wireless networks have also developed overlapping, often complex, relationships. The convenience of wireless access services only heightened the realization of the scarcity of the bandwidth resource upon which it was based. As such, licensed mobile broadband providers sought to secure new bandwidth capacity from wherever possible. Strong arguments were presented to regulatory authorities and policy-makers for the transference of terrestrial broadcast Ultra High-Frequency (UHF) bandwidth capacity to mobile broadband players. Mobile providers also looked to the unlicensed bands, particularly 5GHz, through utilizing so-called off-loading techniques to transfer mobile data from licensed cellular users to unlicensed bands, thus easing the weight of traffic across the former networks (Chen, Yu, Shan, Maaref, and Li, 2016). This was particularly evident in end user devices, for example, as the ability to switch from a licensed mobile broadband to a WiFi network when the latter was available to the user. This kind of activity contributed to the increased levels of traffic on WiFi networks.

The position was further complicated in that licensed cellular communication was underpinned by a different—increasingly competing—technical standard, known as long-term evolution (LTE), developed in the third-generation partnership (3GPP) technical standards-making consortium, rather than the IEEE. LTE reflects and delivers the proprietary features associated with owning tranches of a spectrum as a result of its much more exclusive environment: it is in effect the antithesis of open unlicensed communications. Specifically, 'LTE is designed with the assumption that one operator has exclusive control of a given spectrum; LTE traffic channels are designed to transmit continuously with minimum time gap even in the absence of data traffic' (Babaei, Andreoli-Fang, Pang, and Hamzeh, 2015:338). Cellular communication through LTE operates on a so-called contention-free or schedule-based system, 'designed to transmit persistently as long as there are [data] packets awaiting to be served' (Jeon, Niu, Li, Papathanassiou, and Wu, 2015:2321). This provides it with a better performance capacity compared to WiFi communication. It reflects the world of regulated licensed cellular communication, where the provision of a consistent, high-quality service to customers is considered de rigeur for customer satisfaction and commercial success.

The combined pressures of growing demand, limited capacity, and increasing reliance on the 5GHz band—in particular for off-loading purposes—led licensed mobile broadband players to work on new versions of the LTE technical standard to intensify use of the spectrum in ways more culturally akin to the sector's communications practices and associated cultural values. As shown in the section on the political stream later in this chapter, these interests were key constituents in a complex political stream which exerted influence on the course of development of the 802.11ax standard.

The Policy Stream: 802.11ax, Micro Level Mechanisms, and Intra-organizational Standards-development Pressure. The DensiFi Special Interest Group and Dominance in the Development of 802.11ax

The origins of the 802.11ax standard stretch back to 2012 and 2013 when the IEEE Standing Committee considered submissions about the 802.11ac standard published in March 2013. Products underpinned by 802.11ac had begun to appear as early as 2011, such as WiFi router chipsets and mobile device application processors, followed in 2012 and 2013 by incorporation of 802.11ac functionality into laptops. This provides an indication of the extent to which those involved in creating standards need to see speedy commercial results from the endeavour.

Technical standards-making also involves a seemingly relentless quest for the next generation of technology on which upgraded products can be sold to replace the current generation. In the IEEE, this led to analysis of the just-finalized 802.11ac (which led to the establishment of the Study Group on high-efficiency wireless area networks, known as SG HEW). In line with the regular IEEE standards-making process, the Project Authorization Request (PAR), and Criteria for Standards Development (CSD) documents were produced in order to map out the required work and secure the creation of a Task Group (TG) to undertake the development of the standard. The 802.11 TGax began operations in March 2014. It decided to create a specification framework document (SFD) in line with the 802.11ac process. This is a highly important part of the standards-development process in the IEEE since the SFD is the product of technical development work, proposal to committee, and, ultimately, amendment of draft work in circumstances where a 75% majority approval of any change is required in the balloting process. In March 2016, the technical work on 802.11ax led to the production of a draft specification.

The process leading to the production of a draft specification for a new addition to the family of 802.11 standards is not just significant in technological terms but holds within it important commercial interests and, eventually, consequences. Both these factors combined to create one of the most controversial periods in the history of 802.11 standards-development work, which affected the development of 802.11ax as a policy stream response to the problem stream outlined earlier. The controversy can be characterized as a micro-level mechanism working within the policy stream which took the form of alleged dominance in the standards-development process which had led to the 802.1ax draft specification.

Here, Graham Smith, who had attended the Task Group (TG), argued that technical amendments which he had put forward in the development process had been effectively excluded from what became the draft agreement because of collusive activity between companies, which transgressed IEEE protocols. Smith argued that he had been presenting work on Dynamic Sensitivity Control (DSC) 'with detailed calculations, analysis and technical discussions . . . with over 14 individual submissions (plus revisions) and in the order of 30 other presentations related to DSC'.

He claimed that in 'the spatial reuse ad-hoc group in particular, but also in the TG, every straw poll by me has been voted down' (Stephens, 2016:2). He argued that in September 2015, three presentations related to DSC 'by 108 authors from 16 affiliations . . . 101 authors from 15 affiliations' and, by contrast, '1 author from 1 affiliation' were passed. Smith claimed that he subsequently did more work on DSC and presented 'an independent detailed DSC simulation that showed very positive results' (Stephens, 2016:3) yet this was not passed in a poll. Smith concluded:

> I am convinced that I have been subjected to deliberate exclusion, that my presentations have not been taken on merit, and I believe that a block of members have deliberately agreed among themselves to exclude my idea . . . I therefore can only conclude that this is due to the companies working together in the SIG [Special Interest Group] in TGax which has exercised its influence by reason of superior leverage, strength, and representation to the *exclusion of fair and equitable consideration of other viewpoints*, and in particular, mine (underlining in the original text).' (Stephens, 2016a:3)

An investigation was launched by the 802.11 WG Chair, who is charged with determining whether or not dominance has occurred where, according to the IEEE SASB bylaws, dominance is the exercising of 'authority, leadership, or influence by reason of superior leverage, strength, or representation to the exclusion of fair and equitable consideration of other viewpoints', and to 'hinder the progress of the standards development activity . . . contrary to open and fair participation by all interested parties' (Stephens, 2016b: slide 23).

The investigation involved documentary analysis and interviews by an investigating team from the IEEE. The 802.11 Opening Plenary in July 2016 noted the hope 'that the complete data collection will be completed by the end of the September 2016 meeting' (IEEE P802.11, 2016). In November 2016, a report on the investigation was produced which outlined the process of the investigation and then proceeded to offer a detailed analysis of—and response to—the case made by Smith. The investigation found that 'there have been several SIGs related to 11ax, and specifically that there is one key active SIG currently operating' (Stanley, Montemurro, and Gilb, 2016:5). It specifically noted four SIGs, three of which had low-level activity in the past and were at the time of investigation inactive, alongside, by contrast, a SIG known as DensiFi, whose activity level was high at the time of the investigation.

It noted that the 'SIG referenced in the 11ax complaint . . . is probably DensiFi' (Stanley, Montemurro, and Gilb, 2016:11). DensiFi 'activity appear[ed] to occur largely after Phase 1' of the development of 802.11ax was completed, that is, from January 2015 on. It found the existence of two-tier membership in this SIG: voting/principal membership on motions, held by '7–8 companies' (Stanley, Montemurro, and Gilb, 2016:13) and non-voting expression of opinion through straw polls. The report noted that 'motions determine which presentations or submissions are taken to IEEE802.11ax with SIG members listed as authors'. The report found evidence of a lengthy joining process to the SIG 'limited to 2 companies per 2-month period, effectively preventing individuals and companies from participating in the development of designs and specifications and documents that would be submitted to IEEE 802.11ax' (Stanley, Montemurro, and Gilb, 2016:26), where this 'resulted in a delay such that a company actively working on a product design may not have had access to critical information needed for fundamental design decisions or the ability to participate in the surrounding discussion' (Stanley, Montemurro, and Gilb, 2016:13).

The report went on to provide information about the character of DensiFi. It was noted that there was no information about it publicly available and that it met for three days in advance of each 802.11 meeting 'to discuss and review presentations, and determine documents that are submitted to IEEE 802.11ax' (Stanley, Montemurro, and Gilb, 2016:12). The investigation found that it 'has not entertained proposals from non-members' (Stanley, Montemurro, and Gilb, 2016:12).

However, it also noted that '[a]ll individuals interviewed indicated that they were **not** instructed to vote a certain way by the SIG (bold in original)', though '[s]everal individuals commented that there was an implicit expectation that once a document was agreed in the SIG, that members would act in accordance with the interest of the group and not argue any more about it, and support it in IEEE 802.11' (Stanley, Montemurro, and Gilb, 2016:13). There was strong evidence to suggest that the SIG was successful in securing its preferences in the development of the 802.11ax standard as measured by successful motions. The report noted that 'non-SIG members had 11 failed motions out of a total of 52 (21%). 3 motions proposed by SIG members failed out of a total of 287 motions (1%)' (Stanley, Montemurro, and Gilb, 2016:15).

The investigation concluded that a SIG named DensiFi had been in covert operation over at least two years. DensiFi was made up of a group of prominent ICT players comprising Apple, Broadcom, Cisco, Huawei, Intel, Marvell, MediaTek, Newracom, Qualcomm, Quantenna, Orange, Samsung, Sony, Toshiba, NTT, NTT DoCoMo, and ZTE. The report noted its belief that 'significant discussion on spatial reuse topics, and potential solutions including

DSC occurred within the DensiFi SIG meetings, which Graham did not and could not participate in'. It also agreed that 'the SIG proposals were voted in with less discussion/rationale . . . than Graham Smith's proposal' (Stanley, Montemurro, and Gilb, 2016:25). It concluded that it was 'likely' that 'initially' SIG members agreed 'deliberately' to 'exclude Graham Smith's idea' (Stanley, Montemurro, and Gilb, 2016:25).

The investigating team concluded 'that dominance has occurred through the mechanism of "superior leverage strength or representation" with the effect of excluding viewpoints of non-SIG participants from "fair and equitable consideration" within the 802.11ax Task Group' (Stanley, Montemurro, and Gilb, 2016:27). The grounds for this were the SIG's 'limited openness to interested parties . . . the non-public conduct of the SIG's proceedings and . . . the number of SIG participants as a percentage of the 802.11ax participants' (Stanley, Montemurro, and Gilb, 2016:27). The investigating team recommended that the 802.11 WG Chair 'treat the vote of all individuals affiliated with the DensiFi SG members as a single vote in WG and TG motions and letter ballots related to 802.11ax' (IEEE 802 EC, 2016:1) until the SIG followed IEEE rules or became inactive. Second, it was recommended that 'guidance information regarding the expectation and requirements of individual participation in standards development be developed and provided to 802.11 participants' (IEEE 802 EC, 2016:1).

Reaction to the report was particularly strong from Broadcom, which expressed strong disagreement, objecting to the finding of dominance as well as the 'recommendation that the IEEE 802.11 WG Chair treat the vote of all individuals affiliated with DensiFi SIG members as a single vote'. It argued that the grounds on which the investigating committee drew its conclusions did 'not take into consideration whether or not the DensiFi participants discussed blocking proposals from non-participants, whether or not the participants agreed to vote similarly on certain proposals, or whether or not the participants discussed anything other than technical considerations of their proposals' (Vohra, 2016:1). It was argued that technical considerations should have been 'given far greater weight than the factors relied upon by the investigative team' (Vohra, 2016:1). Broadcom contended that there was 'no evidence indicating that DensiFi SIG participants ever discussed Mr Smith's or any other individuals' proposals, let alone evidence of DensiFi SIG participants discussing the blocking of such proposals' (Vohra, 2016:1). Broadcom was highly critical of the remedy proposed by the investigating team as there was 'no adequate mechanism to determine exactly which individuals voting in the 802.11ax task group are "affiliated" with the DensiFi SIG'. Further, Broadcom noted its strong objection 'to the manner in which evaluation of the investigative report has been conducted', arguing that those parties affected by it should be given appropriate time 'to rebut such findings as well as question and assess the credibility of anonymous sources cited in the investigation' (Vohra, 2016:2). Finally, Broadcom questioned the role of the lead investigator, 802.11 WG vice-chair Dorothy Stanley, a role which Broadcom claimed made her 'an interested party' with 'a vested interest in keeping all discussions related to 802.11 technology within IEEE and thwarting any discussion of such technology outside of IEEE' (Vohra, 2016:2). As a consequence, since the alleged discussions took place outside the IEEE, Broadcom argued that a party external to the organization should have investigated them.

Despite this protest, a motion was put forward at a subsequent IEEE Executive Committee to approve the investigative team's recommendations, moved, and seconded (D'Ambrosia, 2016:14; Gilb, 2016). As a consequence, the vote of all participants affiliated with DensiFi members was to be reduced to one in Working Group and Task Group 802.11ax motions and letter ballots until the SIG ceased to operate. Further clarification from the WG Chair indicated that the individual voting rights of companies would be restored once they had declared their departure from the SIG. This could occur as a result of an authorized letter to the WG Chair declaring either non-affiliation with or withdrawal from DensiFi (Stephens, 2016b:2). As might be expected, this process soon commenced such that, by the middle of December 2016, all members of the consortium had confirmed their departure and, de facto, that DensiFi had been disbanded. By January 2017, the WG Chair was able to report that the measures arising from the 802.11ax dominance matter were no longer in place against any IEEE member (Stephens, 2017).

However, the displeasure at the findings of the investigative team was only matched by the negative reaction to the consequences of their implementation and the effective disbandment of DensiFi. This was manifested by an appeal by Ericsson, Graham Smith, and InterDigital subject to the IEEE SASB Bylaws and Operations Manual. Ericsson holds 146 and InterDigital 282 essential LTE Patents respectively (Mallison, 2015). The complaint sought to challenge 'the fundamentally ineffective remedies supported by the IEEE 802 Executive Committee and the SASB in response to the IEEE-SA's findings' (Ericsson, Smith, and InterDigital, 2017:1). It was also asserted that the 802.11ax draft standard was 'the result of a "corrupted development process"' (Ericsson, Smith, and InterDigital, 2017:2). It was argued that unlike previous SIGs in the 802.11n standard's development, DensiFi was non-public. It was also contended that the latter had 'practically come to supplant actual IEEE-SA technical working groups as the forum for 802.11ax technology development' (Ericsson, Smith, and InterDigital, 2017:A-1). The complainants argued that 'DensiFi participants constitute a majority of voting interests in the 802.11ax working group. . . five of DensiFi's identified members . . . have a combined market share over 93 percent in the integrated circuit market' (Ericsson, Smith, and InterDigital, 2017:A-2).

As a consequence, the appellants called for suspension of all activity related to TGax until effective corrective action was put in place and for a further sixty days after this point (it was claimed that a similar suspension period was put in place in respect of the 802.20 standard) (Ericsson, Smith, and InterDigital, 2017:7). They requested a reduction to one in the vote of DensiFi members and former members, for either six months or until the conclusion of the 802.11ax Draft 1 resolution process. They asked for a re-vote on all motions related to 802.11ax since the time of DensiFi's establishment (Ericsson, Smith, and InterDigital, 2017:2), something, it was claimed, for which there was also precedent in the history of the 802.20 standard (Ericsson, Smith, and InterDigital, 2017:8). It was argued that, in respect of the measures taken regarding the 802.20 standard, 'there is no reason to deviate from the 2006 solution' (Ericsson, Smith, and InterDigital, 2017:A-8). This was necessary action, it was argued, since if left unaddressed, the measures would mean that 'DensiFi's past transgressions would bear no negative consequences and its former members would escape any remedy whatsoever', and would fail to 'discourage future collusive conduct' (Ericsson, Smith, and InterDigital, 2017:4).

It was asserted that the remedies proposed were 'inconsistent with the IEEE's own policies' since 'they do not make it possible for technological proposals that may have been unlawfully excluded to be considered for inclusion in the Standard'. Here it was argued that 'non DensiFi participants have been effectively excluded from fair participation in the development of the Standard' such that 'superior technology and cross-pollination may have been excluded to the detriment of the most innovative and/or efficient technological outcome. It is impossible to know how many valuable or superior technologies, or contributions, were prevented from fair consideration, or discouraged, due to DensiFi's unlawful activities' (Ericsson, Smith, and InterDigital, 2017:7). More broadly, it was argued that the outcome's remedies 'would provide the DensiFi members and future would-be conspirators with the assurance that the IEEE-SA will turn a blind eye to findings of collusion and dominance in the standard setting process'. The appellants also argued that the '[r]emedies expose IEEE-SA to antitrust liability' if unaltered (Ericsson, Smith, and InterDigital, 2017:6). Specifically, it was argued that the 'American National Standards Institute

(ANSI) has accepted the WTO TBT Code of Good Practice, and has since, repeatedly, highlighted the importance of compliance' (Ericsson, Smith, and InterDigital, 2017:A-8) with its elements. It noted that the IEEE-SA is accredited by ANSI and 'adopted a 2011 resolution to adhere to the WTO principles of openness, transparency, impartiality and consensus, effectiveness and relevance' (Ericsson, Smith, and InterDigital, 2017:A-9) and cited IEEE SA's stated commitment to openness in all stages of standards development.

The appellants went further in citing both US and EU antitrust law. They claimed that DensiFi's activity was anti-competitive in respect of the Treaty on the Functioning of the European Union Article 10(1) related to infringement by object and also 'in violation of the EU Guidelines on Horizontal Agreements, Section 7' (Ericsson, Smith, and InterDigital, 2017:A-10). The appellants further referred to Section 1 of the Sherman Act in the United States, related to the act of the deliberate bringing together (or 'rounding up') of parties with an economic interest in the setting of private standards. They also referred to the US Supreme Court's recognition of the potential anticompetitive nature of standards-setting processes and an association's potential responsibility for this, tolerated because of the potential of standards-setting to secure best technology development within open fora (Ericsson, Smith, and InterDigital, 2017:A-10). The appellants called for the IEEE to develop its Antitrust Policy Guidelines to cover the functioning of SIGs in order to 'clarify at what point SIGs engage in the impermissible "execution" of a standard, rather than the permissible setting of the "specifications"' (Ericsson, Smith, and InterDigital, 2017:A-11).

The DensiFi SIG matter played a significant part in the creation by the IEEE's Executive Committee of a new set of instructions, noting specifically the requirement on members to act and vote individually in all IEEE 802 meetings. After the disruption caused by DensiFi dominance, the 802.11ax development process proceeded more straightforwardly. Work on the SFD concluded in May 2017, allowing scope for further development through proposed modifications to the draft amendment. The 802.11ax standard sought to address the key issue of spectral efficiency to ensure higher performance of devices working in an increasingly dense user environment through the use of orthogonal frequency-division multiple access (OFDMA) and 1024 quadrature amplitude modulation (QAM) modulation support to deliver higher signal throughput. It is designed to operate in both the 2.4 and 5 GHz bands, unlike its predecessor 802.11ac standard which operates in the 5Hz band only. The standard also uses Downlink and Uplink multi-user multiple-input and multiple-output (MU MIMO) to allow a user's device to transmit to multiple receivers in both directions at the same time. As such, when finalized, something due to take place at the end of 2019, it is likely to present a (temporary, given the evolutionary nature of the 802.11 standards family) policy solution to the problem stream of developing the performance and efficiency of WiFi in the unlicensed spectrum.

The Political Stream: 802.11ax and Extra-organizational Pressures in Standards Development

The political stream associated with the development of the 802.11ax standard has proven to be as controversial as the internal process associated with developing the standard as a policy solution to the problem of exploiting the 2.4 and 5 GHz bands with the latest WiFi technology. The essence of the political stream concerned the debate on how to secure feasible co-existence between LTE licensed and 802.11 unlicensed spectrum standards. For a detailed analysis of co-existence through the *agenda-setting* lens of the MS framework see Rashid and Simpson (2019).

As noted earlier in this chapter, the 2.4 and 5GHz bands are open to enter environments which have become highly lucrative in commercial terms and upon which a vast consumer public have come to rely to satisfy their daily communication needs. To put matters in context, however, by 2014, it was estimated that more than 22,000 products had been WiFi-certified and that WiFi-based applications had generated as much as US$222 billion in 2013 in the United States alone (OTI, et al., 2015:8–9). WiFi has dominated this part of the spectrum, and its ability to offer consumers the highest quality of communications services possible was seen as imperative. Put another way, the need to address an international consumer public's preferences (or public mood; Zahariadis, 2005) was an important external pressure in the 802.11ax political stream.

Whilst WiFi has undoubtedly dominated the local wireless area network unlicensed spectrum, it has been by no means the only technology to have developed within it, other notable examples being Bluetooth, Zigbee, z-Wave, and Near Field Communication (OTI et al., 2015:10), all of which have co-existed with each other in a relatively harmonious fashion in the unlicensed spectrum. However, the same cannot be said about licensed mobile broadband operators utilizing LTE-based equipment in the unlicensed bands. As noted earlier, the growth of offloading by these licensed mobile providers and their customers into unlicensed bands piqued the interest of the former in exploiting further the available capacity. Offloading is commercially attractive to service providers since the spectrum does not cost anything to enter. It is also attractive to users because switching to WiFi services in environments where they are available either for free or part of a fixed domestic package, for example, allows more extensive online consumption (in terms of duration and capacity) to take place without fear of exhausting the usage allowances which are an often problematic feature of licensed services for smart phone customers.

It is important to note that the 802.11ax standard process was commenced as part of a sequence of standards aimed at improving the quality of communication in the 2.4 and 5GHz bands (confirmed by authors' interviews, 2017). Yet, external factors related to its co-existence with the latest LTE standards soon developed into a second important and highly contentious aspect of the 802.11ax political stream. Here, the 'duty-cycle' technical mechanism incorporated in the LTE-U standard caused alarm in WiFi circles since it did not include the fundamental LBT parameter which underpinned WiFi-based communication. This led to growing concerns over the extent to which LTE-based standards would dominate in the future utilization of unlicensed capacity to the detriment of equipment operating according to different technical parameters, in this case the developing 802.11ax standard. Unlike LBT, the duty-cycle mechanism functioned through a much more possessive approach to spectrum utilization which would allow operators using it to determine when to relinquish control of capacity to other users (Dynamic Spectrum Alliance, 2015). The DSA argued that the mechanism would give control to LTE-U players to determine allocated time to WiFi-based counterparts, and would impact negatively on the real-time voice and video communications opportunity of the latter over WLANs (Dynamic Spectrum Alliance, 11 June, 2015:4). In an increasingly fraught political stream, LTE-U was declared by the Wireless Broadband Alliance, an advocate of WiFi, as insufficient to 'provide for fair use of unlicensed spectrum among all technologies' (Wireless Broadband Alliance, 2015).

The IEEE highlighted the difficulties in defining fairness from the perspective of the different actors involved (IEEE, 2015), arguing that agreement should be reached collaboratively among industry representatives in the concerned self-regulatory SDOs (IEEE, 17 August 2015). It specifically declared its preference for any co-existence to incorporate 'LBT and some sort of exponential back off mechanism' (IEEE, 13 March 2015:6). The IEEE 802 committee, as might be expected, was strongly critical of LTE-U, arguing in respect of the activated duty-cycle that the 'power to make this decision gives LTE-U control over the unlicensed medium and potentially preference for LTE-U systems over 802.11 systems, which is clearly unacceptable for a community resource (unlicensed spectrum) that is supposed to be shared without preference' (IEEE, 13 March 2015:2).

In the eyes of the WiFi community, an alternative LTE standard, License Assisted Access (LAA), developed in the 3GPP consortium, was certainly an improvement in terms of respecting the communications conventions of the unlicensed space yet also became a prominent subject of interest group pressure in the political stream affecting the course of 802.11ax development. LTE-LAA incorporated the fundamental WiFi politeness convention of LBT (Paolini, 2015). Beyond this, however, it was a far from open standard compared to those of the 802.11 family in the important respect that it was designed only for licensed operators using LTE to extend their communications reach more effectively into the 5 GHz band. In so doing, LAA came into direct conflict with WiFi in respect of levels of energy emitted during communication and the consequences of this for other users. The 802.11ax standard, at the time in the throes of a far from unproblematic internal development process within the IEEE, as shown earlier, eventually was affected by the struggle between interests in the political stream, where LTE and WiFi proponents sought to determine parameters of acceptable co-existence in unlicensed bands.

A third important feature of the political stream affecting 802.11ax was government legislative developments, in this case at the EU level. Here, the 2014 Radio Equipment Directive (European Parliament and Council, 2014) mandated the European Telecommunications Standards Institute (ETSI) to update its EN 301 893 wireless standard. The IEEE argued that the directive required 'radio equipment to make efficient use of spectrum and avoid harmful interference' (IEEE, 1 August 2016). Yet the conflict over how this might be achieved in practice between LTE and 802.11 standards remained. An important forum in the process of resolving it was the ETSI Broadband Random Access Networks (BRAN) committee.

A key issue concerned the energy detection or noise thresholds and associated mechanisms incorporated into LTE-LAA and counterpart 802.11 standards, vital in equipment functioning to ensure detection and avoidance of interference. Here, the energy detection (ED) procedure was utilized by LAA devices with a sensitivity threshold of –72 decibel-milliwatts (dBm). By contrast, Wi-Fi standards utilize a two-stage approach which is more sensitive to the presence of energy, with ED at a threshold level of –62 dBm and a so-called Preamble Detection (PD) procedure operating at an energy threshold of –82 dBm. This raised two potential problems for WiFi devices. First, less sensitive LTE devices might be more likely to produce signals colliding—and causing interference—with other users in the unlicensed spectrum band. There was a concern that this could be exacerbated with more intensive LTE device usage of the unlicensed spectrum. Second, and stemming from this, 802.11 devices, not least those that would be underpinned by the 802.11ax standard, might suffer a degree of crowding out from the 5GHz band as LTE devices increased their presence.

In the EN 301 893 revision process evolving in ETSI BRAN, the IEEE 802.11 argument was that the architects of LTE-LAA in 3GPP should adopt a more sensitive ED threshold or incorporate a dual-mode PD and ED function. In response, 3GPP argued that the IEEE in its work on 802.11ax should adopt an ED-only mechanism at the threshold of –72 dBm utilized by LTE-LAA. This external conflict in the political stream became more intense as the need to ensure that the 802.11ax standard, at this point still in development, would be covered by the revised EN 301 893, once completed. Ensuring this occurred entailed commercial as well as techno-functional importance given consumer reliance on - and preference for - the best quality WiFi technology in their personal electronic devices. In an attempt to overcome the continuing lack of consensus, negotiation between 3GPP and IEEE 802 took place via formal liaison statement exchanges, as well as participation in workshops organized with the purpose of finding consensus on key areas of fair coexistence. In 2017, a revised version of the EN 301 893 standard set out two options for accessing unlicensed spectrum by radio equipment: a single ED threshold and a dual threshold with PD. This amounted to a holding position in which the stalemate between the licensed cellular and WiFi communities was evident since the latter adhered to the dual-ED level approach and the former the single approach.

In a further development in 2018, ETSI BRAN produced four potential technical solutions to the ED impasse, one of which might be adopted in the revised EN 301 893 standard. As the discussions on securing an agreement on energy levels proceeded, it became clear that co-existence had become the key issue in the 802.11ax political stream, resolution of which was likely to involve the continued acceptance of the differing approaches to ED advocated by the LTE and 802.11 sides of the debate. More specifically, the inclusion of IEEE 802.11ax in the revised ETSI standard was considered by WiFi interests as germane to securing any forward movement on co-existence. A key problem, however, was that 802.11ax was still under development at the time. As discussions on the matter in ETSI intensified in the first half of 2018, the differences between the parties continued to be evident. Yet, unexpectedly, an individual proposal emerged from a participant with interest in developing LTE in the unlicensed spectrum which advocated that, for the purpose of revising the ETSI standard, participants consider 802.11ax (which was due to be completed in the following year, 2019) as a 'market reality' (authors' interview, 2018).

This was accepted by ETSI BRAN as a way to exit the EN 301 893 standard-making deadlock. It also allowed the political stream to be combined with the problem and policy streams in progressing the development of the 802.11ax standard. The political stream compromise was in line with the previous version of the EN 301 893 standard with which 802.11 functionality was compliant, yet provided a normative twist in recognizing an—at that point—incomplete standard in 802.11ax. Additionally, it was agreed that a 'real-world' performance testing process for LTE-LAA and 802.11ax, conducted by ETSI and WiFi Alliance members, needed to be developed (IEEE 802.11 Coexistence SC, 2018) to provide evidence of the quality of their co-existence with each other in unlicensed bands.

Conclusion

The history of the development of the 802.11ax standard is illustrative of the high stakes involved in LAN wireless communication and the complexities which can ensue both internally to SDOs in the policy stream but also in external environments characterized by the political stream. The controversial nature of the early development of 802.11ax within the IEEE illustrates the strong desire of powerful commercial players to secure and pursue their interests in lucrative standards-development processes. As the basis for future product sales and service functionality in an unlicensed spectrum environment still predominantly occupied by WiFi, 802.11ax was characterized by special interest group behaviour considered unacceptable by the IEEE's investigating team. This investigation, and its outcome, was contested, yet the case of the DensiFi SIG and dominance in the 802.11ax process also provides evidence of a detailed intra-organizational process and a clear degree of openness and candour in the standards-making world of the IEEE.

In a very different way, though no less controversially, the 802.11ax standard shows how the products of private sector driven international standards-making processes can come into direct conflict with one another, creating external pressure on the development of standards like 802.11ax. As a corollary, the pursuit and maintenance of co-existence between products and services underpinned by standards with oppositional functionality, occupying a finite and highly sought-after communications space, becomes imperative though challenging to realize. The case illustrates how the desire to capitalize on consumer appetite to switch as seamlessly as possible between licensed wireless and WiFi services can ultimately drive the motivation to pursue combined understanding and practices of co-existence between standards whose technologies, in the shape of LTE-LAA and 802.11ax, ultimately deliver services which are de facto highly similar to users, the vast majority of whom are unconcerned about the technical standards underpinning their speedy and reliable access to the Internet, undertaken wirelessly, and often on the move. The 802.11ax standard, the latest in a series of WiFi standards developed over the best part of thirty years, though at times deeply controversial, ultimately displays evidence of how problem, policy, and political streams in the private sector standards-development world can be successfully combined, in what is the latest phase in the continuing evolution of the unlicensed spectrum environment in which WiFi has developed with such commercial success.

7

The Do Not Track Standard

The Failure of Self-regulation and the Politics of Contestation

This chapter will focus on an issue—online tracking—that has implications in terms of privacy rights and, more specifically, the right to choose whether or not to be tracked when browsing the Internet. The proliferation and persistence of online tracking of consumer behaviour through various technologies for the purpose of collecting information and constructing targeted ads based on the profiling of the interests and preferences of users, at surface level, seems relatively harmless. As pointed out by certain analysts, it can result in 'an interesting ad or a tempting commercial offer' (Kamara and Costa, 2016:14). From a marketing, advertising, and publishing perspective, such activity is crucial to revenue generation (profit) and to the provision of online content. However, the broader impact and implications of such activity is much more wide-ranging, in particular given that much of it is not transparent and that it can result in the violation of a plethora of individual rights including (mis) use of personal data, practices that lead to discrimination in the price of services and products, a narrowing (and bias) of the information accessible online (filter bubble) to users, as well as the collection of personal data that can be used, without explicit consent, in the pursuit of both profit and national security. Beyond this, such tracking practices can cause insecurity and anxiety among those individuals that express concern at the loss of personal control and privacy online. The Facebook/Cambridge Analytica scandal of 2018 is testament to this.

It is argued that whilst not focusing on policy-makers and public policy directly, the Multiple Streams (MS) framework (Kingdon, 1984, 2011; Zahariadis, 2007; Ackrill, Kay, and Zahariadis, 2013; Cairney and Jones, 2016; Cairney and Zahariadis, 2016) provides a useful conceptual starting point for understanding decision-making in complex, multi-actor, standards-developing organizations (SDOs). The MS lens, more specifically, allows the identification of windows of opportunity for change within the three streams and can be supplemented further with the identification

Global Standard-Setting in Internet Governance. Alison Harcourt, George Christou and Seamus Simpson, Oxford University Press (2020).
DOI: 10.1093/law-oso/9780198841524.001.0001

of specific internal (micro) and exogenous (macro) mechanisms (Falleti and Lynch, 2009; Gerring, 2010) in SDOs that enable and constrain change. This chapter assesses the Do Not Track (DNT) initiative and voluntary (open) standard as one response to tracking behaviour that has sought to ensure that users have a choice on whether they are tracked or not. It focuses on the extent to which stakeholders have influenced its conception, character, and implementation in the World Wide Web Consortium (W3C). Moreover, it is argued that whilst a series of indicators (rather than a focusing event) opened a window of opportunity in the problem stream to address the issue of online tracking, conflict in the policy and politics streams has hindered the ability of policy advocates to construct a (universally) consensual standard that is usable, credible, and enforceable.

To this end, the first section of this chapter pays attention to the appearance of the problem. The chapter turns next to the policy stream and the W3C's contesting of and deliberations on the DNT standard. The chapter then analyses the politics strand and how the changing political and legal climate constrains or enables actors within the W3C to alter the DNT standard and achieve consensus regarding its implementation.

The Problem Stream: Trends, a Window of Opportunity, and Initial Coupling

The story of the emergence of the DNT standard dates back to 2007 when pro-privacy advocates (policy entrepreneurs, in this case) such as the World Privacy Forum (WPF), Consumer Watchdog, the Center for Democracy and Technology (CDT), and the Electronic Frontier Foundation (EFF) requested that the US Federal Trade Commission (FTC)—which has oversight authority for consumer protection and corporate competition in the United States—create a DNT list for advertising online. Politically DNT was not just backed by national governments but also by prominent EU Commissioners such as Vivian Redding and Neelie Kroes who led on the evolution of the EU's Information and Communications Technologies/Information Technology (ICT/IS) agenda for many years. Their view of and support for DNT was not backed by all their senior staff—for instance, a British member of Kroes' Cabinet opposed it—but nevertheless the W3C campaign to design and create a DNT standard was backed by the Commission. Indeed, personal information management, protection by default, and the right to privacy was embedded within the EU's Data Protection Directive (1995) and its replacement, the General Data Protection

Regulation (2016), in particular in the specific protocol on the Right to Erasure (the Right to Be Forgotten).

The problem, from a public interest perspective, was that self-regulatory efforts such as the Network Advertising Initiative (NIA) had failed to protect the privacy of consumers in the collection, use, maintenance, and disclosure of personal and behavioural information. In other words, consumer privacy was being increasingly threatened in new ways that many consumers were largely unaware of, through the expansion of behavioural tracking and targeting (Centre for Democracy and Technology, 2007). Whilst this initiative provided inspiration for privacy and security researcher Christopher Soghoian and Mozilla engineer Sid Stamm to develop a DNT header, the first attempt to take this idea or proposal to a standards body—the Internet Engineering Task Force (IETF) rather than the W3C—was met with very little interest and was thus not submitted as a proposal.

Window of Opportunity

DNT did not effectively acquire real political and policy momentum until the media and the Wall Street Journal in particular reported and drew public attention 'to the sophisticated and non-transparent tactics used to track users online' (Toner, Eckersley, and Reitman, 2015). The FTC announced that it would investigate online tracking in December 2009. In July 2010, the Chairman of the FTC, Jon Leibowitz, in the course of his oral testimony to Senate Commerce Committee on online privacy, endorsed the idea of a DNT list being explored further (Soghoian, 2011).

The FTC effectively called for the creation of—among several proposed options from different interests—a DNT mechanism that (i) would apply across industry to all types of tracking; (ii) be easily located, easy to use, effective, and enforceable; and (iii) would allow consumers to stop collection of behavioural data gathered across sites and not just the serving of targeted ads (FTC 2010, 2012). The FTC also proposed US Bills for the purpose of establishing a DNT mechanism, but many (e.g. the Do Not Track Me Online Act of 2011) have not been enacted. By 2011, Sid Stamm was already implementing a form of DNT in the Firefox browser and Microsoft had set the Do Not Track value (1) as the default setting in Internet Explorer 10—the other values being (0) meaning the user wishes to be tracked, and (null) meaning the user has not set a preference. Whilst such a setting was obviously the most privacy friendly (thus supported by pro-privacy groups) as it meant users had to actively change the setting to (0), advertisers and publishers argued that they were being denied the opportunity to engage with users interested in targeted advertising (and thus to make additional profit) as many users lacked the knowledge or were very often reluctant to change default settings once set.

From an MS perspective, this chapter explains the appearance of the DNT issue on the W3C agenda through an initial coupling of the problem and politics streams—the former in terms of opening a window of opportunity for finding a solution to the issue of tracking and the latter in relation to providing regulatory and political momentum for this task through the W3C. Practically, there were exogenous and internal mechanisms that contributed to W3C putting the DNT standard on their agenda. First, because it seemed like the natural home for it given it was the standards body for Hypertext Markup Language (HTML) (even though they are not the only body involved in Hypertext Transfer Protocol (HTTP)), and second, and perhaps more symbolically, because as founder of the World Wide Web and leader of the W3C, Tim Berners-Lee put his weight behind the concerns voiced in relation to online tracking.

Third, many individuals who had worked on the Platform for Privacy Preferences (P3P) inside the W3C were not happy with the outcome and saw DNT as an opportunity to put things right (on P3P see Faith Cranor, 2012). Furthermore, the diversity that existed in initial efforts to implement DNT as a browser header was also a central reason that the W3C undertook the task of developing a consensual standard that could be universally applied. The W3C had taken an interest in DNT in February 2011 and held a workshop where all stakeholders offered their perspectives (W3C, 2011). Following discussions at the workshop, the Tracking Protection Working Group (TPWG) was established in September 2011, Aleecia McDonald and Matthias Schunter co-chairing, the former working on the compliance aspect and the latter on the technical aspect. The compliance side, perhaps predictably, was more contentious as it posed questions of what it would mean for companies (in particular in large technology and advertising): what should and could be tracked (or what was not being tracked)? What is the minimum companies would need to do to show that they were honouring DNT?

Contention also surrounded the role of Aleecia McDonald as co-chair; some industry and advertising representatives saw her as representative of civil society (a 'privacy person'); others, from civil society organizations (CSOs), thought she should represent their interests (she was half-time at Stanford

University and also worked half-time at Mozilla). She saw her main task as arbitrator between the many different interests represented in the TPWG.[1] For her, the public interest would be served by forging consensus on a workable standard, which proved to be an enormously difficult task. It was also acknowledged early on in the process that it was unlikely to be anything but a slow and protracted task given the sheer number of participants within the working group—initially over 100 people—and the diversity of interests represented. Whilst the main civil society representation came from the EFF (with, atypically, three representatives), the CDT and the FPF were also present, and invited 'public' experts were also brought in (Walter van Holst, for example). This said, it was clear that larger corporations and advertising representatives made up the majority of the TPWG, and that a richer pool of resources to draw upon allowed them to participate in proceedings more consistently. For instance, in the initial formation of the TPWG, Microsoft deployed five representatives to work on the DNT standard.

The Policy Stream: Contestation and Consolidation of Diversity

There were several key issues at stake that caused substantive delay in making progress to Candidate Recommendation[2] in the TPWG; that is, achieving consensus on a DNT standard that would accommodate different but not necessarily all concerned interests. It can be argued that, despite one of the W3C's underlying reasons for addressing this problem through the creation of a TPWG being that of trying to create a unified positon among diverse interests, there is little evidence that stakeholder interaction within the group resulted in any socialization towards collective consensus. The institutional mechanism of the W3C in this case, it can be argued, was heavily populated and dominated by the interests of large tech companies and the advertising industry. It was effectively the actions of those actors that, through both their lobbying against further privacy regulation relating to tracking in the politics stream despite public dissatisfaction with privacy rights exacerbated by events such as the Facebook/ Cambridge Analytica scandal, and their refusal to countenance alternative interpretations of DNT inside the TPWG (see later in this chapter), meant that

[1] See: https://www.w3.org/2000/09/dbwg/details?group=49311&public=1&order=org.

[2] For the different stages of progression for a standard in the W3C, see: https://www.w3.org/Consortium/Process/Process-19991111/tr.html#RecsCR.

the DNT standard would result in the (re-) enforcement of divergent positions on DNT, and ultimately, user preferences.

There existed a contentious debate on the meaning of DNT among stakeholders that was also obvious at the W3C workshop held in 2011 (W3C, 2011). The FTC (and indeed CSOs') position was that data collection should be limited, but many companies argued that this could not possibly be done as data was needed for security (e.g. companies such as PayPal), whilst pro-privacy advocates insisted it was not necessary. The question of data collection limitation was therefore contentious between the different stakeholders, and was made even more difficult to resolve given regulatory differences between the European Union (collection/use based on proportionality) and the United States (where the FTC argued for limitation but also granted permitted uses where required). There were also additional difficulties (practical, in some instances, such as finding a compatible time for meetings) in terms of incorporating the Asian perspective and this did not bode well when attempting to construct a worldwide applicable DNT standard. More specifically, there was no agreement or consensus on what DNT meant. Stakeholders, broadly speaking, took the following positions within the TPWG discussions:

- Advertisers/Publishers: the Interactive Advertising Bureau (IAB) and the Digital Advertising Alliance (DAA)—which together represented the majority of the advertising industry—advocated a self-regulatory approach which took DNT to simply mean collection could continue but users could opt-out of targeted ads (e.g. Yahoo included). The group also saw the high-profile defection of the DAA in 2013. DAA members said they would develop their own solution that would allow consumers to opt-out of targeted advertising (it was the same solution that was offered prior to the TPWG commencing its work).
- Google: sought to remove identification of individual data through replacing unique identifiers with same string identifiers. Thus, they argued for a single string op-out that effectively meant the data collected could not be used in the same way (to target individuals), as individual data could not be distinguished. Such a proposal meant tracking and collection of data would still be possible, however. Google, effectively even if more innovatively, sat in the same group (as did Facebook) as the advertising coalition (both are members of the IAB).
- CSOs/Privacy advocates (and pro-privacy companies such as Mozilla): supported privacy by default, and/or providing users with genuine choice that would be honoured and implemented by companies

(with a minority of stakeholders taking DNT with a literal meaning; that is, the need to stop tracking altogether).

Differences in terms of distinctions between and interpretations of legal concepts were also widespread and caused delay. For example, there were disagreements about potential exemptions and the implications of this for consent and use of user data,[3] on the distinction between first parties and outsourcing service providers acting as first parties (i.e. what technically and legally constitutes a first party—what are the technical limitations and how far can they be solved?), as well as issues relating to first- and third-party compliance with DNT, where first-party content is technically dependent on third-party content, but the latter is neither trusted nor compliant, providing little genuine choice for the user if they want to enjoy the full content.[4] There were also additional issues relating to the compatibility between EU and US laws, and the assumptions embedded within those with regard to, for example, US legal concepts such as first/third party and how far they were compatible with EU legal (equivalent) concepts such as data controller/data processer. Moreover, there were issues surrounding technically incorporating EU legal requirements (see also Article 29 Working Party, 2014) such as *explicit* informed consent into the DNT standard as well as ensuring interoperability in terms of implementation of the standard within the European Union and across different legal regimes.[5]

Whilst Microsoft was proactive in setting DNT as a default preference for users in Internet Explorer (IE) 10,[6] after two years of implementation and non-compliance by companies that did not respect the preference, it announced that future versions of IE would not implement it by default. Instead it offered a DNT option in its IE browser and gave users clear instruction on how they could make a choice on whether they wanted to be tracked or not. In effect, it reacted to the arguments made by the W3C that such a setting would not reflect a deliberate choice on the part of the user, and importantly that it provided an excuse to ignore the DNT signal, particularly to those that claimed they did not comply precisely because it was not clear that it reflected user preference (Kamara and Kosta, 2016:7). Advertisers and publishers and certain

[3] See: http://www.w3.org/2011/tracking-protection/track/issues/147.

[4] See: http://www.w3.org/2011/tracking-protection/track/issues/184.

[5] See: http://www.w3.org/2011/tracking-protection/track/issues/189 and http://www.w3.org/2011/tracking-protection/track/issues/183.

[6] Mozilla also launched a private browsing version of its web browser with the DNT function automatically turned on. AVG, the antivirus company, similarly provided users with an opt-in to DNT by default. https://www.avg.com/en-gb/privacy.

technology companies that rely on tracking consumer behaviour for revenue and profit continue to ignore the DNT signal at worst or pay lip-service to it through acknowledging user preference but not respecting user choice.

The influential self-regulating DAA, effectively the largest representatives of online advertising companies, defected from the TPWG in September 2013 making it virtually impossible to achieve the aim of a universally (adoptable) applicable standard. They developed their own self-regulatory solution called Ad Choices that would allow consumers to opt-out of targeted advertising but that allows companies to continue tracking them. Twitter, which was one of the leaders in supporting DNT in its inception, abandoned the idea in 2017 in favour of more granular privacy controls precisely because an industry-wide standard approach had not materialized (Vaughan-Nichols, 2017). Another major disincentive to continue with a DNT standard in the W3C (at least this was the interpretation of many advertisers) was the success rate that Mozilla had enjoyed when initially implementing DNT (17% adoption rate) in their Firefox browser; an expectation, it seems, for many in the advertising world that joined the TPWG was that its adoption would be much lower, thus having a much less critical impact on revenue and profit.

Advocacy groups such as Consumer Watchdog (based in California) abandoned the DNT standard because of its voluntary nature; for them it was not strong enough, it was laden with loopholes and there was no effective enforcement mechanism. The EFF, a prominent civil society member of the W3C TPWG, took matters into its own hands, releasing a final version of a free plugin called Privacy Badger for Firefox and Chrome browsers (August 2015). It acts as enforcer when a user turns on the DNT preference, scanning websites to determine if the publisher has agreed to honour the privacy request. If it cannot find a policy, it scans for third-party scripts that appear to be tracking, and blocks them. Exogenous mechanisms—in this case, technological development in the form of privacy software—have, in the case of the EFF at least, not helped to enable consensus on DNT within the TPWG but rather consolidate difference in terms of the diversity of positions on tracking and enforcing DNT. Other companies have also developed anti-tracking and DNT technological solutions that provide users with real choice in relation to tracking including Safari, Firefox, and new market players such as Brave Browser and Better. In this sense, DNT policy is being developed by different stakeholders, independently of the progress of the DNT standard but which, in different ways, offers resistance to online tracking.

To this end, the EFF has also developed (working with Disconnect and AdBlock) a DNT policy—or code of conduct—positioned as a compromise proposal that would allow users to avoid tracking and publishers to continue tracking to gather ad revenue. In short, if publishers honoured users' preference not to be tracked, they would get an EFF seal of approval or certification that this was the case, and publishers could subsequently give users an option to disable ad-blocking software allowing them to show ads. The major advertisers, ad networks, and technology companies have not signed up to this policy, but smaller tech firms such as Medium (a blogging platform) and Mixpanel (an analytics firm) have, and it is hoped that others will follow once the benefits (privacy certification, user trust, competitive advantage) become more obvious in the medium term (Dwoskin, 2015).

The Politics Stream: Timing and Legal Reinforcement?

To what extent can legal frameworks help to underpin the DNT standard produced and enforce a more robust form of privacy for users? How has the broader political climate and public mood, in relation to privacy, affected the ability of actors in the W3C TPWG to forge change towards consensus?

This chapter has already alluded to the US FTC and EU Commission's advocacy of DNT as part of a tightening up of consumer privacy rights, in particular after 2009. This said, in the United States in particular, the FTC's support for some sort of regulatory approach to pervasive online tracking has consistently been weakened by market mechanisms. The power of technology and advertising actors was witnessed in relation to their interpretation of FTC principles, successful resistance against the Consumer Privacy Bill of Rights released by the Obama Administration (and other bills introduced by this Administration addressing digital privacy) and the positions they were able to introduce and sustain in the multi-stakeholder W3C due to the imbalance in membership which favoured resource-rich industry players. All of this, weakened the ability of pro-privacy advocates (in particular CSOs) to create an ideational consensus towards a universal DNT standard that provided users with a genuine choice on tracking. DNT would only be acceptable to major technology companies and advertising groups if: (i) they could continue to collect, use, and monetize data; (ii) DNT was not set to default in browsers (in particular as this meant loss of profit); and (iii) they were allowed to continue self-regulating.

Despite political pressure, the technology and advertising industry was able to curtail regulation in the name of innovation and, simultaneously, constrain any further discussions on DNT in the W3C. Since the Trump Administration came to power, political support has waned regarding privacy with resistance from the US Senate in March 2017.[7] More broadly, under Trump, it seems there is little appetite to advance any US law that enhances online user privacy over US national security (see also Brookman, 2015). More specifically, whilst the Facebook/Cambridge Analytical scandal in 2018 can be considered as a focusing event that (re-) opened a window of opportunity for those advocating a more robust regulatory approach to protecting user choice and privacy, many doubt that the practice of the tech companies and advertising industry will change in relation to DNT; indeed, these companies strongly oppose any new privacy regulation or legislation. Thus, given the political climate in the United States (California perhaps the exception here), it is unlikely that such a window will result in coupling the politics and problem streams and substantive change in the attitudes towards—and contestation related to—enhanced regulation, DNT, and enhancing privacy rights more broadly.

What then, of the laws that do exist in relation to enhancing privacy? One of the first laws to support the DNT effort was established in California. The California Online Privacy Protection Act (CalOPPA) requires the operator of a website or online service to display a privacy policy that discloses the categories of personally identifiable information (PII) collected through the site, the categories of third-party persons or entities with whom the operator may share that PII and other content (Hengesbaugh and de la Lama, 2013). The CalOPPA was amended in 2013 (California AB 370, 2013) and introduced two new content obligations for the site operator to:

> Disclose how the operator responds to web browser's Do-Not-Track signals or other mechanisms that provide consumers the ability to exercise choice regarding the collection of PII about an individual consumer's online activities over time and across third-party websites or online services, if the operator engages in that collection, and
>
> Disclose whether other parties may collect personally identifiable information about an individual consumer's online activities over time and across different websites when a consumer uses the operator's website or service.

CalOPPA further specifies that an operator may satisfy the first requirement above by providing 'a clear and conspicuous hyperlink in the operator's privacy policy to an online location containing a description, including the effects, of any program or protocol the operator follows that offers the consumer that choice' (Ibid). CalOPPA does not enforce DNT, but rather is a disclosure law

[7] See: https://www.eff.org/deeplinks/2017/03/senate-puts-isp-profits-over-your-privacy.

that requires transparency (notice) about the behaviour of operators regarding DNT. For some, 'transparency policies about the response to DNT are a necessary preliminary ingredient for a future strong DNT regulatory enforcement mechanism' (Kamara and Kosta, 2016:9). For others, and as confirmed through interviews, whilst it may offer users more of an informed choice on tracking it is does not prohibit tracking altogether, there have been major problems with its enforceability, and even when imposed, fines have been derisory thus not providing companies with incentives to comply with the law. Moreover, under the CalOPPA, whilst sites need to say whether or not they honour the user's DNT preference, it 'doesn't go far enough to incentivize companies to *actually* honour DNT' just to merely disclose how they address the preference. Thus, it does not in any practical way empower users. The updated California Data Privacy Protection Act (also known as AB 375) is a piece of consumer privacy legislation which passed into California law on 28 June 2018 (but will not come in to full effect until 1 January 2020), and for some is the US equivalent to the General Data Protection Regulation (GDPR) (European Union 2016). This brings with it more stringent privacy protection for users which is likely to have significant implications for user privacy rights, including in relation to their right to explicit choice on tracking.

Within Europe, the EU's GDPR and the ePrivacy Regulation (ePR) (at the time of writing still being deliberated over)) provide the user with additional safeguards and rights with regards to objecting to tracking for marketing purposes, including profiling (O'Neill, 2018. The ePR moves away from simply the 'right of consent' offered by its predecessor the ePrivacy Directive (European Union 2009, Article 5(3))[8] with regard to the installation and use of cookies (see Kosta, 2013, on regulation of cookies in Europe), to require that users' ' informed, *specific and unambiguous*' (emphasis added) consent be obtained before websites store or use personal identifiers such as tracking cookies. It also stipulates the right of users to revoke any such consent at any point in time. In terms of the ePD, a central concern related to the applicability of the DNT preference (the Tracking Preference Expression (TPE)) within the EU, in particular has been in meeting its requirements in Article 5(3) for valid and informed consent (Kamara and Kosta, 2016:10–12). The ePR has essentially been updated to remedy this and align with the higher levels of protection in the EU GDPR, including the high level of fines for companies that are found not to comply (4% of turnover). Indeed, one might argue that the GDPR has become

[8] Introduced in 2009 via the 2004 Citizens' Rights Directive, moving away from the 'right to refuse'.

a global standard with those wanting to do business in the EU having to upgrade their laws and business practices.

There are, of course, certain exceptions—indeed, limitations to the rights within the GDPR—that is, users (Art. 21) cannot object to profiling when it is based on the performance of a contract (Art. 6(1)(b)) or the consent of the data subject (Art. 6(1)(a))—which amounts to a considerable number of cases where profiling does occur. Similarly, in relation to Article 22, there is an exception to the right not to be subject to a decision based solely on automated processing when such a decision is necessary for entering or the performance of a contract between the controller and the data subject. There is a clear intention within the EU to strengthen the right of users in terms of their ability to object to the tracking, profiling, and use of their personal data. As noted by the European Data Protection Supervisor (EDPS) in a review of the e-Privacy Directive (EDPS, 2016; EDPS 2017; see also European Commission 2015a):

> Consent should be genuine, offering a freely given choice to users, as required under the GDPR. There should be no more 'cookie walls'. Beyond a clear set of exceptions (such as first party analytics), no communications should be subject to tracking and monitoring without freely given consent, whether by cookies, device-fingerprinting, or other technological means. Users must also have user-friendly and effective mechanisms to provide and revoke their consent within the browser (or other software or operating system).

However, questions have been raised by the EU's Article 29 Working Party about the potential alignment of the DNT standard with EU legislation in relation to terminology, the definition of tracking, the purpose (scope) of tracking, explicit informed consent, default browser settings and tracking, and enforceability (Article 29 Working Party, 2014; see Kamara and Kosta, 2016, for discussion of these challenges). In short, whilst the legislation within the EU certainly provides a legal basis and platform for future, strong DNT regulatory mechanisms of enforcement, it is also clear that further thought must be given to strengthening and aligning such mechanisms with the DNT standard (and vice versa)—that is, to ensure that there is a uniform and clear terminology, that the technical and legal provisions are compatible, and that users' right to choose on tracking is clear and transparent. Furthermore, there is also the question of how far such a standard is compatible with the law in other regions and countries of the world.

Conclusion

The DNT standard has been the subject of much contestation since its inception. The idea for the standard emerged from a push by policy entrepreneurs (consumer organizations and privacy advocates) to ensure user privacy, following trends in persisting tracking. Once identified, the tracking problem then found its way on to the W3C agenda through its coupling with political momentum, in particular the media and the push by the FTC to establish consensus and find a solution to it that was agreeable to all stakeholders. However, the W3C institutional mechanism worked against both consensus and solution. Indeed, market and institutional mechanisms, coupled with technological evolution in the form of privacy software, meant that the perceptions of and positions on how to resolve the tracking problem were further entrenched in the W3C. Ideational clashes rather than collective learning became the norm with change difficult to achieve towards overall consensus on a universal DNT standard. Socialization between members of the TPWG did not lead to the creation of a solution that brought stakeholders closer together. Rather, it ossified positions developed externally and favoured corporate solutions (tech and advertising industry) over those advocated by pro-privacy groups; self-regulation and implicit choice won over explicit choice and the unambiguous right to choose by the user in relation to tracking.

To this end, key stakeholders defected from the W3C TPWG developing DNT or they lost faith in it. Pro-privacy consumer organizations and advocates have developed their own technological solutions and policy to ensure that consumers have a choice on whether they are tracked or not, but the W3C DNT standard will not result in a universal (all stakeholders) acceptance of and compliance with the standard. In addition, the voluntary nature of the standard has meant that major opposing interests (major technology companies and advertisers) have been able effectively to ignore the DNT standard. This said, the recoupling of the problem and politics streams through embedding similar standards in the law (with regards to transparency, user consent, and explicit user choice on tracking) have emerged, in particular through the updated California Data Privacy Protection Act and the EU GDPR and (the forthcoming at the time of writing) ePrivacy Regulation. This, alongside the Facebook/Cambridge Analytica scandal, has, for privacy advocates in particular, created a momentum once again for change. However, and despite the public mood and appetite for unambiguous consent in relation to how their data is used, there is little sign that technology and advertising groups in the United States will compromise on either the voluntary DNT standard or indeed any (US-wide) legislation that seeks to regulate privacy rights.

If advertisers and publishers continue to ignore the DNT standard—or at least the need legally for *explicit* consent of users in terms of their personal data being tracked—then they are likely to end up with much more severe technological solutions that block advertising in its entirety (e.g. Privacy Badger and other ad blockers) and more severe compliance and enforcement mechanisms including substantive financial penalties for those companies that do not respect the users 'default' right to explicit consent to protect their personal data, if that is their choice. This is an outcome that will potentially impact both user experience and company revenues in particular if there is alignment between the DNT standard with and across jurisdictions. If privacy and respect (compliance) for users' choice not to be tracked is implemented, it might very well lead to a loss of some revenue in the short run, but such losses would be offset in the medium to long term through increased trust on the part of consumers in those companies that take their privacy seriously, therefore providing a competitive advantage over others that do not. Moreover, by respecting those users that do not want to be tracked, technology companies and advertisers can reduce costs and improve targeted ads for those who are indifferent to tracking.

8

Technical Standards, Dynamic Spectrum Access, and Competing Spectrum Policy Interests in the TV White Space Environment

The seemingly inexorable growth in mobile communications has led to strong pressures to change the way in which scarce spectrum space has been managed and used historically. What has become a major communication policy challenge has recently given rise to a recognizable problem stream in a hitherto underexplored part of radio spectrum, so-called television white spaces (TVWS). The specific issue has been an attempt to open up to more extensive use the narrow 'spectrum buffer zones' sitting between bands allocated historically for terrestrial TV broadcasting. The spaces in question have been occupied by companies from the Programming Making and Special Events (PMSE) industries. These incumbents have faced new entrants wishing to provide broadband Internet services on a commercial basis. The latter has up until now involved provision of 'standard' Internet access but TVWS has also been viewed as a potentially important way of delivering IoT services in the future. A third important party in an evolving picture has been terrestrial broadcast service providers, which have been concerned about the consequences of more intensively utilized TVWS, not least in respect to potential signal interference.

Initiatives to reform TVWS use, led by the United States and the United Kingdom, are underpinned by vital technical standards. Here, the goal has been to introduce computer-mediated Dynamic Spectrum Access (DSA)— ideally through unlicensed entry—to implement tiered priority use of the wireless environment. DSA instructs devices wishing to operate in TVWS on the frequencies that can be used and for how long (Temperton, 2015) and can be regarded as a form of implemented governance through technology. TVWS users from broadcast network operations and PMSE have first- and second-tier priority over use of the spectrum. New entrant TV white space players wishing to provide Internet access services are third in the usage 'pecking order'.

Global Standard-Setting in Internet Governance. Alison Harcourt, George Christou and Seamus Simpson, Oxford University Press (2020).
DOI: 10.1093/law-oso/9780198841524.001.0001

The specific approach this chapter adopts is to use the MS framework applied to the decision-making phase of policy (after Zahariadis, 1992) to provide an understanding of the role which technical standards-making has played in policy efforts to open up TVWS. The matter has proven controversial and contestation continues. It highlights the contingent nature of technical standards-making in processes of policy change. The chapter shows how the core problem of creating an acceptable and efficacious technical system of DSA in order to make better use of scarce spectrum capacity has been addressed readily through technical standards-making, principally in the Institute of Electrical and Electronics Engineers (IEEE). It argues that whilst the potential policy solutions to underpin the realization of DSA were thus created, coupling with a highly contentious political stream has been challenging and has up until now delivered only relatively embryonic socio-economic benefits. In line with the ideas of Zahariadis (2005), the chapter illustrates how the TVWS political stream consists of assertive and often strongly opposed interests from outside the technical standards-making process, namely terrestrial broadcasters, PMSE providers, television audiences, and TVWS new entrant Internet service providers. The political stream is also characterized by the policy activities and preferences of key states, in our case the United States and the United Kingdom.

The chapter shows that whilst technical standards-making has played a vital role in creating the conditions for the development of equipment to exploit new opportunities to attempt to use TVWS more intensively in an historically carefully ordered and distinctly closed spectrum, much more is needed to secure deployment and development of the new communications services underpinned by such developments. Technical standards development in the private standards-developing organization (SDO) context of the IEEE has been affected by external actor preferences in the political stream in this complex picture. The chapter shows how opportunities for policy change are constrained by conflict when powerful incumbent interests from commercial, public service, and civil society quarters express scepticism—and even out-and-out resistance—to such change. In these circumstances, often characterized paradoxically by both turbulence and sluggishness, market growth and scale can be difficult to realize.

The topic of TVWS provides a particularly interesting illustration of the often fractious nature of policy problems in spectrum use and their interplay in the political stream, despite the clear (technical) policy stream solutions generated by significant SDO activity. The implications of this for the provision not only of standard Internet access services to promote rural connectivity but also into the future for the development of IoT services are explored in the chapter.

The Problem Stream: How to Intensify the Use of TVWS Spectrum

Digital terrestrial TV has been provided through the use of the Ultra High Frequency (UHF) spectrum divided into multiple channels across territorial regions. To prevent signal interference between regions, spectrum buffer zones, commonly known as white spaces, were created. Here 'channels are allocated to geographical areas in such a way that adjacent areas use different frequencies and have relatively large guard bands between them, so as to avoid interference between transmissions' (Brydon, 2013:2). These served the additional purpose of providing what became a well-established and highly closed-use environment for broadcast players from the programming making and special events industry—and sound and acoustic services delivery, in particular.

However, as the demand for spectrum capacity has burgeoned, TV white spaces have widely become viewed as un(der)-utilized. It has been estimated that up to 100 Mbps services may be possible through techniques which combine the capacity of multiple available channels (Nominet, 2018) (see later this chapter). The UHF band of terrestrial TV is particularly valuable since signals can travel further and through and around physical objects. Importantly, antennas can be small enough to fit on mobile equipment (Brydon, 2013). The US pioneered the pursuit of policy change to intensify the use of TVWS. In Europe, the United Kingdom was the first EU state to develop a detailed TVWS use expansion policy agenda.

However, potentially broadening the use of TVWS had a daunting associated problem stream characterized in significant part by technical challenges. The most pressing concerned technical interference. Since the original raison d'etre of TVWS itself was to avoid signal interference, allowing communicative activity with them was understandably viewed with caution. The technical feasibility of delivering interference-free increased levels of communication in TVWS was also layered with issues of incumbency and priority use which manifested themselves strongly in the political stream (see later in this chapter). The pressing need for more spectrum availability also sat in contrast with the distinct technical limitations of TVWS capacity. Generally, urban environments, being densely populated, are likely to be less conducive to the introduction of services based on TVWS than less populated rural environments.

Yet, the potential development of TVWS pointed to the capacity to address two urgent issues each with very different normative and practical features underpinning them. The first concerns socio-economic development in communications where the most prominent instances are the possibility of providing online connectivity to those in rural and other non-densely populated locations and, by contrast, provision of affordable access to basic universal service level online services in selected urban environments (authors' interview, 2018) based on socio-economic characteristics and technical feasibility.

Rather differently, the second area where change in TVWS use has been considered important is in providing machine-to-machine (M2M) online services in the context of the IoT. In 2011, Wilmington, North Carolina, used TVWS to provide IoT services related to public lighting and water-level monitoring (Gilpin, 2014) in the context of so-called smart cities (Gibbs, 2013). TVWS has three to five times more range than WiFi, for example. In a different way, latency in IoT devices can be as long as 5 to 20 minutes and still be effective where other devices need as little as 20 milliseconds (authors' interview, 2018). Efforts to resolve the problem and develop the potential of TVWS through technological progress were small scale initially. In 2005, Microsoft created the first TVWS radio which was deployed in a US campus setting (authors' interview, 2018). Yet, it has been estimated that as many as 3.9 billion people globally are unable to afford fixed broadband access (Gilpin, 2014). These and similar considerations lay behind significant work which took place in the international private sector driven standards-setting environment of the IEEE.

The TVWS Policy Stream: Intensifying the Use of TVWS Through Dynamic Spectrum Access

The use of TVWS is attractive since its frequencies have good propagation properties (Mueck and Noguet, 2011:1). The IEEE 802 committee soon became the source of the main international technical standards that could efficaciously underpin the development of TVWS use. The 802.22 standard, which was launched after a landmark US Federal Communications Commission (FCC) Notice of Proposed Rule Making in 2004 (see later in this chapter), is focused on distances of up to 100 kilometres and is thus a wireless regional area network (RAN) standard. The 802.15.4m standard related to low-power wireless personal area network communication in TVWS was originally developed in 2003. In 2014, the IEEE finalized the 802.11af standard to underpin short-range local area network (LAN) TVWS transmission of up to one kilometre.

In 2014, the IEEE produced the IEEE 802.19.1 for TVWS coexistence between different television band networks (Filin et al., 2018). The European Computer Manufacturer's Association (ECMA) International then developed the ECMA-392 standard for Media Access Control (MAC) and the physical layer for operation in TVWS. In 2016, the IEEE produced the 1900.7 standard for white space dynamic spectrum access radio systems. Cognitive radio adds to so-called software-defined radio 'to learn from its operating environment and adapt to statistical variations in the input stimuli for efficient resource utilization' (Masonta, Makgotlho, and Mekuria, 2012:2). The Dynamic Spectrum Management which results requires sensing, decision-taking, sharing, and mobility to be created in a single technical system.

Both fixed and mobile devices were envisaged as being able to operate in TVWS. The 802.22 standard uses cognitive radio techniques and has been described as the 'first worldwide effort to define a novel wireless air interface standard based on cognitive radios (CRs) . . . to operate in the TV broadcast bands provided no harmful interference is caused to incumbent services (e.g. TV receivers)' (Cordeiro, Challapali, and Shankar, 2005). In respect of the IoT, the sister 802.22.1 standard aims to protect low-powered devices against interference in TVWS. As part of this significant suite of technical standards-making, the IEEE has aimed at ensuring that equipment using each standard can co-exist in the part of the communication space where they overlap. So-called opportunistic access, which emerged from research on cognitive radio techniques, has been defined as a way of providing access to spectrum for devices with second tier access rights in a way that ensures no harmful interference with first tier priority users (Struzak and Wiecek, 2013) . The 802.22 standard was principally designed for effective so-called incumbent detection through spectrum sensing. The development of TVWS has been a policy priority in the United States and the United Kingdom. Their respective communications regulators, the FCC and Ofcom, have opted for the introduction of geo-location databases as the main technical means for detecting the incumbent users (Masonta, Makgotlho, and Mekuria, 2012:3), holding tiered priority access rights to the spectrum. Geo-location techniques work when the primary user of the spectrum is registered with a database such that any cognitive radio wishing to communicate must detect its location and find free channel space. The 802.22 standard was 'the first wireless air interface standard focused on the development of CR based WRAN physical (PHY) and medium access control (MAC) layers for operation in TVWS' whose physical layer uses orthogonal frequency-division multiple access (OFDMA) technology (Masonta, Makgotlho, and Mekuria, 2012:5). The 802.11af standard uses the

TV frequency bands located in several different white space sub-bands from 54 MHz to 790 MHz (authors' interview, 2018). The wavelength of 11af is on average ten times greater than the 5GHz 11ac standard used for WiFi-based communication.

The most salient legislative development in the EU context was the radio and telecommunications terminal equipment (R&TTE) Directive which was important for setting the context for the functioning of TVWS services in Europe and thus affected the United Kingdom. This began with the development of a TVWS database and specification for database-to-database communication. The work was undertaken by the European Telecommunications Standards Institute (ETSI) (through its Reconfigurable Radio Systems Technical Committee). The ETSI EN 301 598 wireless standard articulates 'the European . . . standard for TV white space devices' (Dynamic Spectrum Alliance, 2014:4). The DSA's Model Rules and Regulations for the Use of Television White Spaces (Dynamic Spectrum Alliance, 2017:10) state that devices 'must comply with the emission limits specified in ETSI EN 301 598 v1.1.1' and possible subsequent versions of it.

A key issue in the development of TVWS was the accommodation of mobile communication since this would require dynamic handover and access techniques to be developed. The European Conference of Postal and Telecommunications administrations (CEPT) studied TVWS use by so-called secondary devices (CEPT WG SE43). According to Masonta, Makgotlho, and Mekuria, the International Telecommunication Union (ITU) has undertaken work in the field of cognitive radio technology standardization through two working parties (ITU-R WP 1B and WP 5A) which have considered 'the concepts and regulatory measures required to introduce CR' (Masonta, Makgotlho, and Mekuria, 2012:4).

In the United States, the FCC in its Notice of Proposed Rule Making related to the 600 MHz TVWS band, articulated the principle of 'Use it or Share it'. This was heavily endorsed by the DSA (a diverse group of commercial and non-commercial actors which advocates for the growth of dynamic spectrum access underpinned services) which even argued for its application to the broader spectrum policy environment. So doing, it was argued would, among other things, discourage spectrum purchase for non-use and future speculative selling on. For the DSA, 'a spectrum policy that balances licensed and unlicensed approaches will maximise innovation and investment' (Dynamic Spectrum Alliance, 2014 :3). To date, the United States, United Kingdom, Canada, and Singapore have led the drive to create rules for licence-exempt access to TVWS. At the global level, the 2012 International Telecommunication

Union World Radio Conference noted that its regulatory framework was able to accommodate software defined and cognitive radio systems (Dynamic Spectrum Alliance, 2015). Masonta, Makgotlho, and Mekuria, present the interesting idea that any cognitive radio in a DSA system needs to be both 'spectrum and policy agile', where the former involves 'operating over a broad range of frequency spectrum' and the latter an awareness of the 'constraints under which it operates' (2012: 2). These constraints in TVWS policy development are very clearly illustrated through analysis of the political stream which developed in the respective cases of the United States and the United Kingdom and are the focus of the remainder of the chapter.

The Contested TVWS Political Stream: Policy Development and Its Challenges in the United States

The potential expansion in the use of TVWS stretches back to a 2002 Notice of Inquiry from the FCC which sought to launch a debate on the 'feasibility of allowing unlicensed devices to operate in TV broadcast spectrum at locations and times when spectrum is not being used, and on the technical requirements that would be necessary to ensure that such devices do not cause interference to authorized services operating within the TV broadcast bands' (Federal Communications Commission, 2002:1). Here, the FCC, steering strongly for the more intensive utilization of TVWS, made the now familiar points about spectrum availability in the spaces between licensed TV transmission bands. It also pointed to what was then the prospect of the switchover from analogue to digital television services and the extra capacity this would provide.

Importantly, it made some early reference to the development of technology to utilize parts of the spare spectrum dynamically, which can be interpreted as an entrepreneurial attempt to couple the TVWS problem and policy streams where 'advances in computer technology mean that it should be possible to design equipment that would monitor the spectrum to detect frequencies already in use and ensure that transmissions only occur on open frequencies . . . a device could have the capability of "knowing" where it is, and could use information obtained from a database to determine whether there are any licensed operations in its vicinity'. It also argued that 'equipment can be designed that is frequency agile, with the capability of changing operating frequencies or bandwidth as needed to avoid interference to licensed users' (Federal Communications Commission, 2002:6).

This important policy intervention was followed in 2004 by a Notice of Proposed Rulemaking which declared the intention 'to allow unlicensed radio transmitters to operate in the broadcast television spectrum at locations where that spectrum is not being used' (Federal Communications Commission, 2004:1). The FCC argued that this move would promote more efficient spectrum use, allow the development and deployment of new innovative services, and would even be beneficial to the broadcasting industry, since it could provide it with the opportunity to provide new broadband–broadcast converged services.

However, the Commission was also cognizant of—and acknowledged clearly—the differences of opinion on the possible move to develop unlicensed TVWS use. Here, 'parties representing the interests of manufacturers and users of unlicensed devices generally support[ed] this approach, while those representing the interests of the current users of the TV broadcast spectrum, both primary and secondary, express[ed] concern about potential interference from such new unlicensed operations' (Federal Communications Commission, 2004:4). This assertion succinctly highlighted the character of the TVWS political stream. The input of transmission and reception equipment manufacturing companies was important in the development of the US TVWS sector. Carlson Wireless, along with Dell, GE, Microsoft, and Google, formed the White Space Coalition. Carlson offered to manufacture necessary hardware to exploit the business opportunity available in developing the concept of cognitive radio; the FCC incentivized approval with experimental licenses for trials of cognitive radio (authors' interview, 2018).

In the 2004 and subsequent 2006 Notices (Federal Communications Commission, 2006), the FCC elaborated on advocates for the introduction of unlicensed TVWS services. These included corporate players like Intel, Microsoft, and Motorola; wireless Internet Service Providers' representative bodies such as the Consumer Electronics Association and the Information Technology Industry Council; the wireless standards-making committee IEEE 802; and parties such as the Wi-Fi Alliance and the New America Foundation.

The DSA expressed a position put forward by the Wi-Fi Alliance which argued for the creation of four MHz channels for unlicensed spectrum use assembled from a combination of the so-called guard bands between mobile and broadcast transmissions, the 'duplex gap' between radio bands for licensed mobile communication, Channel 37, which it was argued could in certain cases be used as a co-existence channel between unlicensed users and 'medical telemetry and radio astronomy' devices and co-existence with wireless microphone users in the channels reserved for the latter. It was also contended that on a 'market specific basis, unlicensed devices should be permitted to continue using unassigned channels in the re-packed UHF spectrum under the FCC's TV white space rules'. (Dynamic Spectrum Alliance, 2014:5). It declared its opposition to any expansion in the rights of licensed microphone operators in the 600 MHz guard band and duplex gap and argued that the FCC 'should also make available for unlicensed use any unassigned spectrum remaining after the auction of the LTE uplink and downlink channels' (Dynamic Spectrum Alliance, 2014: 2).

Despite this, these proposed changes to the use of TVWS in the United States faced significant opposition in the political stream. For example, the FCC noted that broadcasters represented by Association for Maximum Service Television, Inc., the National Association of Broadcasters, and the Association of Public Television Stations had expressed opposition to any such move. Parties from commercial mobile radio services and private land mobile radio services, as well as wireless microphone manufacturers, expressed concerns about interference from any potential change in usage of white spaces. The FCC noted fears about interference to broadcast signals received by consumers but also to TV transceivers, and the view that it would be appropriate to postpone the introduction of TVWS services until the move to digital broadcasting had been completed and to introduce such services on a licensed, rather than an unlicensed, basis (Federal Communications Commission, 2006:5).

Despite this, the FCC argued that it 'disagree[ed] with broadcasters regarding the potential of technical solutions for ensuring that unlicensed devices only use vacant channel space' concluding that '[w]hile we understand the concerns of broadcasters and other existing authorized users of the TV bands about the possibility of new interference, we believe that with appropriate safeguards it would be possible to allow unlicensed operation in the TV bands' (Federal Communications Commission, 2004:7–8) without causing interference.

In 2006, in another reference to a policy solution to the TVWS problem, the FCC noted the technical standards-making activity of the IEEE 802.22 Working Group, formed in the light of the Commission's 2004 Notice, to create a technical standard for unlicensed communication in TV white spaces. It noted that the group had 'broad participation from those in the TV broadcast sector and the public safety community who are licensed users of TV spectrum, as well as from chip vendors, wireless equipment suppliers, and firms from other countries (Federal Communications Commission 2006:6).

In 2008, the FCC released a Second Report and Order which adopted rules to allow unlicensed radio transmitters to operate in TVWS (Federal

Communications Commission, 2008) which was followed in 2010 by a Second Memorandum Opinion and Order finalizing the rules. Stressing its view of the importance of developing TVWS, it noted its innovative qualities with many economic and social benefits where the 'potential uses of this spectrum are limited only by the imagination' (Federal Communications Commission, 2010a:2). The development of the US TVWS policy has been underpinned by so-called incentive auctions where broadcast licence holders agreed to relinquish at least part of their spectrum in return for a share of the proceedings from its sale. This process has proven controversial. The FCC noted that it had received as many as seventeen petitions which 'collectively request numerous changes in the rules for TV bands devices' (Federal Communications Commission, 2010a:3). The IEEE was aware of the importance of its 802.11af standard for development of the TVWS and its 802.18 committee lobbied to try to ensure that the TVWS incumbents would not be disadvantaged by changes in band allocations (authors' interview, 2017).

In response to concerns raised with it, the FCC made modifications in respect of protecting incumbent services (principally TV and wireless microphones); placing further requirements on devices that would operate in TVWS; measures to ensure security of communications between TVWS devices and databases; and measures to protect TV stations in neighbouring Canada and Mexico. The FCC signalled that manufacturing of TVWS devices could soon commence but cautioned that it would 'closely oversee the introduction of these devices to the market . . . [taking] . . . whatever actions may be necessary to avoid, and if necessary correct, any harmful interference that may occur' (Federal Communications Commission, 2010a:8).

An important part of the US TVWS system is the use of DSA databases. The Commission's Second Report and Order stipulated the requirement for devices operating in TVWS bands to register in, and use, a database to check for available channels in their location. As part of the standards underpinned policy stream solution to the TVWS problematic, the databases would provide a 'privately owned and operated service that unlicensed TV bands devices must contact to obtain information on channel availability at the locations where they are operated and, in the case of fixed devices, to register their operation at those locations' (Federal Communications Commission, 2010a:40). In addition, the database was required to hold information on authorized services in the band, the service paths of broadcast auxiliary point-to-point facilities, the regions served private land mobile radio service and commercial mobile radio service (PLMRS/CMRS) operations on channels 14–20, regions served by the Offshore Radiotelephone service, and locations of cableheads and low-power reception sites outside 'the protected contours of the TV stations whose signals they receive' (Federal Communications Commission, 2010a:41). Despite some concerns raised in the light of the Second Order, the Commission upheld its 'decision to allow the designation of multiple database administrators and . . . [to] . . . rely on market forces to shape the structure of the database administration functions and service offerings' (Federal Communications Commission, 2010b:44). Information required in the database was mandated to be made publicly available. Spectrum Bridge arose as an important player in the TVWS database industry. In June 2013, the FCC awarded Google the right to operate a national US TVWS dynamic spectrum access database. The service it provides is free to use for individual (though not business) customers.

Despite these important policy developments in the creation of TVWS services in the United States which suggest a combining of the problem and policy streams and the highly contested political stream, the sector can still be regarded as being at a relatively early stage in its development (authors' interview, 2018). There have been some notable, though relatively small-scale, examples of TVWS roll-out projects in rural areas, such as the Homework Gap project which provided an online connection to 200 homes in Virginia. The broader roll out of TVWS services in urban areas is considered a much greater challenge yet to be realized (authors' interview, 2018).

The Contested TVWS Political Stream: UK TVWS Developments

The United Kingdom has been a pioneer in intensifying the use of TVWS through DSA. However, like the US case, the political stream has proven to be highly contested. The strategic significance of TVWS was illustrated in 2011 in the United Kingdom through the Cambridge white space trial funded by Microsoft, British Telecom (BT), the BBC, and Nokia. In a clear effort to drive forward the UK TVWS policy agenda in October 2013, the UK communications regulator, Ofcom, launched a consultation based on a raft of TVWS proposals, derived as a result of a series of thirty-five teleconference consultations with key interests in TVWS in the United Kingdom, which occurred between September 2011 and April 2013 (BBC, 2013). This period provides a clear illustration of the fractious and contested nature of the TVWS political stream in the United Kingdom. Here, issues of spectrum rights, resource reallocation, and future functional co-existence rubbed shoulders uneasily. The debate comprised articulations of the value of public interest present in the status quo in terms of consumption of digital terrestrial television, claims about the need to capitalize on policy solutions provided by technical standards development and associated technologies to optimize new communications service possibilities, and concerns about the maintenance of incumbent communications service provider rights to (co-) exist in increasingly close proximity to new entrant TVWS Internet service operators.

The TVWS political stream was noteworthy because of the presence of vociferous interest groups. Responses to the Ofcom consultation illustrate clearly their different perspectives on the ways in which TVWS should be developed in the United Kingdom. A first set of interests—Digital Terrestrial Television (DTT) advocates—provided a response which ranged from full-scale scepticism to, at best, cautious acceptance of change. Arguably the most powerful DTT player, the BBC, recognized that 'successful implementation of TVWS technology could enhance the service that we provide and open up new avenues of accessing content for our audience'. Yet, rather differently, it also contended that 'these positive outcomes can only be realised if existing and highly valued BBC services are not negatively impacted by a flawed introduction of TVWS services'. From this basis, it went on to criticize Ofcom, arguing that the proposals were 'significantly different to the position agreed in Ofcom's technical working group, and as such carry an elevated risk to TV and PMSE services' (BBC, 2013:2) where 'a number of sensible positions appear to have been abandoned' (BBC, 2013:4–5). It accused Ofcom of not adhering to the best practice of undertaking an impact assessment of its proposals. Furthermore, it argued that 'some critical points of policy as they relate to TVWS may have been decided or pre-empted without consultation with stakeholders'. The result was that the proposals 'appear to be giving priority spectrum access to licence-exempt services over lawful licensed services' (BBC, 2013:5), something which it argued was unique in UK spectrum management.

The BBC was keen to contrast the treatment of DTT stakeholders with those from the licensed mobile broadband business using the Long Term Evolution standard on the 800 MHz band. Clearly linked to the broader politics of spectrum policy change in the United Kingdom, where licensed mobile broadband operators had been awarded increased capacity at the expense of digital broadcasters in the 800 MHz UHF band, it noted that whilst 'over two years of BBC measurement data on DTT services' was 'not considered', then by contrast 'tellingly, protection has been tightened for IMT services in LTE-800 band' (BBC, 2013:6). Of particular concern was future interference to DTT viewers utilizing internal aerials to receive signals. Claiming that proposals it had made to protect these users had been ignored, coupled with Ofcom's move to allow increased power levels for TVWS devices operating internally, the BBC asserted frankly that 'it is not tenable to adopt a policy position which results in as many as 25% of all TVs facing significantly elevated risks of interference from a licence-exempt service' (BBC, 2013:7).

The BBC further expressed concerns about the growth of IoT services using TVWS. It argued that '[r]eceiver performance issues, identified through two years of DTT measurements by the BBC, have not been acknowledged'. Furthermore, it was claimed that an 'optimistic' set of so-called protection ratios had been proposed in the consultation document on the basis of the Weightless IoT Special Interest Group which had developed an IoT M2M standard 'that had been hard wired to generate a benign test signal' which was 'in no way representative of real M2M or similar WSD applications'. Additionally, the BBC claimed that 'data from DUK [Digital UK] has been truncated with the result that WSD have been favoured' with 'a high risk of interference if WSDs are widely deployed as proposed by the M2M industry'. Finally, it was claimed that the 'protection of PMSE has also been relaxed' and consensus reached in the European Communications Committee, which Ofcom led the pursuit of, had been ignored' (BBC, 2013:8).

The British Entertainment Industry Group (BEIRG), which represented the interests of the PMSE community, was equally forthright in its criticism of Ofcom's proposals. It argued that TVWS deployment had 'the potential to severely compromise PMSE's operating environment' since the 'conclusions of five years' of work by ITU, CEPT and Ofcom's Technical Working Group with regard to the protection levels required for PMSE have been largely ignored in favour of the WSD community's projections'. It presented a stark picture where damage to PMSE providers 'would negatively affect practically every UK citizen and consumer, and damage the creative industries that contribute £36 billion annually to the UK economy and provide 1.5 million jobs' (British Entertainment Industry Radio Group, 2013:2). The BEIRG also challenged the value of the TVWS industry to the United Kingdom, not least given uncertainty about the future of the 700 MHz band in the United Kingdom (see Rashid and Simpson, 2019b), asking pointedly 'What does Ofcom perceive to be the *realistic* projections for revenue and jobs from the introduction of WSD?' (emphasis in original) (British Entertainment Industry Radio Group, 2013). The key concern for the PMSE lobby was signal interference. Dismayed at Ofcom's use of the term 'harmful interference', the BEIRG noted unequivocally that 'as far as the PMSE industry is concerned, *any interference is harmful*' and would prove 'disastrous' for broadcasters and other events providers (emphasis in original) (British Entertainment Industry Radio Group, 2013:3–4).

The BEIRG also questioned specifically the economic argument for WSD, noting its concerns over whether white space services would 'provide sufficient value to taxpayers, when weighed up against expenditure on proposed database management by Ofcom, the threat of damaging interference and potential compensation costs to disrupted industries' (British Entertainment Industry Radio Group, 2013:4). In particular, it noted that 'were producers to be blamed for any WSD interference, a situation would very quickly arise where shows could no longer get insurance to hold a production. This would be disastrous for the industry, which would be all but halted as a consequence' (British Entertainment Industry Radio Group, 2013:5).

The Digital TV Group declared that several of its members were 'keen to exploit opportunities with TVWS' (Digital TV Group, 2013:1). Yet, it also raised concerns about technical protection ratios for DTT receivers to prevent interference. It also echoed concern about the use of only a Weightless device to test for this which it argued could cause 'the corresponding protection ratio to be set too low with no adequate protection for a real-world WSD deployment' (Digital TV Group, 2013:2).

The UK's civil society body, the Voice of the Listener and Viewer, representing terrestrial TV audiences, raised the matter of the need for TVWS equipment and associated use practices to be compliant with the ETSI EN 301 598 technical standard. Like its DTV advocates, it was 'concerned about the short timescales proposed and that too cursory an evaluation of the effects of WSDs may result in irreversible impairment of DTT and PMSE services to licence payers' (Voice of the Listener and Viewer, 2013:4). The VLV noted its broader concern about change in the use of the UHF spectrum and its possible effects on viewers and listeners. In respect of potential interference to internal TV aerials, it argued that 'licence fee payers that are adversely affected will not take kindly to the impairment or removal of that service and the cost incurred in finding a remedy' (Voice of the Listener and Viewer, 2013:5). The VLV also expressed concern over what it saw as a difference in regulatory approach for DTV and PMSE in the proposed regime. Specifically, whilst Ofcom was responsible for managing the relationship between DTV and WSDs, the latter relationship looked set to be a form of self-regulation yet was a situation in which 'interference is potentially more harmful and where calculations have to be made far more often in more complex environments' (Voice of the Listener and Viewer, 2013:8).

The VLV also raised concerns about the economic viability of any WSD compliance regime and went as far as to argue for consideration of a possible shared access fee or similar contribution to ensuring a well-resourced conformance scheme (Voice of the Listener and Viewer, 2013:9). Overall, it held 'deep concerns that the very nature of the proposed use of WSDs is covert as far as DTT licensees are concerned and so a robust policing process MUST be in place' (Voice of the Listener and Viewer, 2013:7). It also noted the so-called downstream revenue losses that could be accrued from sub-standard performances. It argued for a more cautious approach than that proposed by Ofcom involving the use of exclusion zones, low-power transmission, detailed technical checks conducted in non-lab based settings, and the acknowledgment of the mobile and temporary nature of broadcast and entertainment events. The BEIRG was particularly critical of Ofcom's proposal to update the WS database every 15 minutes which it regarded as 'too long a time for PMSE operators to wait on occasions when instantaneous access to interference-free spectrum is required (Voice of the Listener and Viewer, 2013:5).

The digitaluk alliance, owned by the BBC, ITV, Channel 4, and Arquiva, was unsurprisingly concerned about the Ofcom's TVWS proposals. Its Freeview DTV service was at that point 'the sole television platform in more than 10 million homes' in the United Kingdom (digitaluk, 2013:2). It expressed 'concerns about the adequacy of protection for DTT services from TV white space' (digitaluk, 2013:3). It expressed concerns about the risks entailed in Ofcom's proposed strategy, noting that there was 'no precedent for knowingly permitting interference from a licence exempt service into a licensed service' (digitaluk, 2013:5). It argued that 'the likely response times to address complaints of loss of DTT reception are unlikely to be sufficiently short to guarantee consumers continued enjoyment of a particular programme' (digitaluk, 2013:5). In a direct criticism of Ofcom but also a direct challenge to the TVWS sector, it argued that Ofcom had changed technical parameters agreed in consultations conducted prior to the release of its proposals 'to the potential detriment of DTT reception and yet this is still represented as a cautious approach' (digitaluk, 2013:6).

It also expressed strong support for PMSE users noting its long-standing 'synergistic' relationship with the latter, stating pointedly that the 'PMSE/DTT sharing arrangement is licensed; well managed and is significantly more cautious than that proposed for TV white space' (digitaluk, 2013:6). Sony, a manufacturer of television sets, supported these concerns, noting the specific point that 'allowing an increased level of interference in stronger DTT signal areas may cause interference to consumers with less than ideal amplifier/cable screening setups which are currently working perfectly well, due to the strong DTT signal and low interference levels' (Sony, 2014:1).

In contrast to these perspectives on the change to TVWS use in the United Kingdom, a second broad set of interests active in the political stream comprised supporters of TVWS development. Here, there was a diverse range of players with different potential roles in a more intensively used TVWS environment.

Rival broadcasters to DTT providers, linked to the broader politics of change in spectrum policy related to the DTT bands, agitated for the opening up of the UHF spectrum which they considered to be an unnecessarily restricted environment for the provision of communications services. The UK licensed mobile broadband company Vodafone argued forthrightly that the 'approach to the protection of DTT reception that Ofcom has proposed... is totally at odds with the approach that Ofcom proposes for TVWS in this consultation' (Vodafone, 2013:2). As an organization interested in exploiting the 700 MHz band vacated by incumbent DTT providers under recent spectrum policy change in the United Kingdom, Vodafone noted an apparent difference in view by Ofcom of the robustness of DTT reception interference between users of the 700 MHz band and would be entrants into TVWS. Specifically, for Vodafone, the approach taken in the case of the 700 MHz band appeared more cautious whereas

> given that TVWS devices operate on a secondary basis, the criteria for coexistence might be expected to be more stringent than for 700 MHz band terminals ... [where] ... [u]nduly stringent requirements for 700 MHz terminals in Europe will threaten global roaming and economies of scale which will be to the detriment of citizens and consumers. (Vodafone, 2013:7)

Elsewhere, it argued that Ofcom's technical report in the consultation documentation was 'based on incorrect assumptions on the performance of 800 MHz band terminals' which led to the concern that Ofcom might need 'to review whether some restriction of operation of WSDs is needed on channel 59, in order to avoid harmful interference to 800 MHz band terminals' (Vodafone, 2013:3), that is, essential mobile phones. In respect of the TVWS database, it criticized Ofcom's move to define the WSD database algorithm from the outset. Vodafone was also critical of what it speculated might have been unnecessarily complex proposals that had been developed too quickly. The Weightless consortium, with a strong interest in the development of IoT-based machine-to-machine communication argued that 'much previous regulation on sharing and interference has been overly conservative (Weightless Special Interest Group, 2013:1). In contrast to the concerns expressed by DTV and PMSE interests, it agreed 'strongly with the approach of only protecting rooftop reception and not indoor reception'.

Sky was forthright in arguing that '[o]ver-protection for certain groups based on the cascading of worst case protection elements should be avoided, when it is clear that doing so would unnecessarily hinder the launch of this dynamic innovation' (Sky, 2013:1). It praised Ofcom's proposals, arguing that its 'practical approach should not be undermined by endless theoretical argument about worst case scenarios as it has in the past' (Sky, 2013:2). It even went on to argue that certain of Ofcom's proposals were 'too restrictive and risk stifling the potential of new white space applications' (Sky, 2013:3). It targeted the two main incumbents in the UHF band by urging Ofcom to ensure that 'channels are only unavailable to white space devices for as long as they are being actively used by PSME applications' and arguing that 'the DTT platform (which acts as the biggest constraint to white space devices in the UHF band) should be subject to ongoing appraisal with regard to its spectrum efficiency' (Sky, 2013:3).

From the telecommunications sector, in contrast to DTT supporters, BT argued that 'the general approach proposed by Ofcom is reasonable and appropriate for ensuring a low probability of harmful interference to DTT services' (BT, 2013:2) and that its approach provided 'a good balance between precision, complexity and practicality' (BT, 2013:5). From the perspective of incumbent Internet services providers, the telecommunications services provider EE argued that 'Ofcom's approach to ensuring a low probability of harmful interference to 4G services is insufficiently robust' (EE, 2014:1).

Google, a significant player in TVWS database technology development in the United States highlighted its support for the use of databases to allow co-existence to be managed. It was critical of Ofcom's approach, arguing that by 'precluding TV white space operations where there is a mere possibility of interference to a receiver, Ofcom has proposed an overprotective approach that limits the utilization of white space, even when there may be no interference' (Google, 2015:7). The M2M communications company Neul argued forthrightly that 'the availability of TV White Space is at the limits of what is viable for commercial deployments, especially in urban and sub-urban region' (Neul Ltd, 2013:1). Directly noting its desire to capitalise on the IoT potential of TVWS, Neul contended that the power limit to be imposed by Ofcom on TVWS providers was too stringent. It criticized calculations made on the coverage area of white space master devices which it argued would create generic, and thus unnecessarily constraining, operating parameters. The techUK body, representing 270,000 digital technology equipment and services companies, criticized the consultation for not addressing the issue of amplified overload which it argued had caused interference during the rollout of 4G mobile communications services. It argued that the 'approach of Ofcom involves the use of a number of approximations and assumptions that make it challenging to assess the final impact on reception quality for the viewer' (techUK, 2013:5). It argued that Ofcom's framework was unlikely to create interference to PMSE services. The Wi-Fi Alliance argued that Ofcom's building penetration loss estimate 'of 7bB seems to be overly conservative' (Wi-Fi Alliance, 2013:2). Similarly, it argued that 'the absolute nature of the power levels for non-geolocated slave devices could result in significant overprotection and diminished use of available spectrum' (Wi-Fi Alliance, 2013:3). It argued against a proposed reservation of channel 60 and instead argued that 'Ofcom should just state the out-of-band emission limit that a WSD shall comply with' or limit the power that a WSD could emit in the channel (Wi-Fi Alliance, 2013:5).

This illustration of the fractious political stream in the debate on UK TVWS policy clearly shows the opposition from broadcasting interests to change but also the challenges to these interests from licensed mobile players, as well as those wishing to exploit TVWS specifically. As noted above, 2012–13 was also a period in which the bandwidth held by terrestrial broadcasters in the United Kingdom came under scrutiny in respect of possible transfers to the licensed mobile sector (see Rashid and Simpson, 2019b). This arguably more important matter may, for DTT providers, have led to a softening of their stance on the development of unlicensed TVWS services (authors' interview, 2018). The PMSE community proved highly effective in articulating their interests, not least their desire for certainty about their future position (authors' interview, 2018). In the process of negotiating the TVWS framework, it has been claimed that incumbents, unlike new TVWS providers, had 'unfettered access to the regulator' (authors' interview, 2018).

Yet, in February 2015, the United Kingdom took the decision to implement TVWS services through a system of dynamic spectrum management, making specific mention of its intention that TVWS spectrum would 'form a major part of what is becoming known as the "Internet of Things"' (Ofcom, 2015:2). The UK system is licensed, though the ambition is to develop an unlicensed open access environment populated by equipment which is able to communicate its location in an 'autonomous' fashion. In the first instance, TVWS devices were required to communicate with the DSA database in return for information on available communication channels and the permissible power level at which to transmit. Periodic contact with the database is required to ensure that no interference with other users, particularly those of the first- (terrestrial broadcasters) and second-tiers (PMSE providers), occurs (Nominet, 2018).

Ofcom created 'a central control system . . . supported by a number of independent database operators' (Brydon, 2013:5–6). By the end of 2015, four database operators were registered to provide TVWS DSA access services: Fairspectrum Oy, Nominet UK, Sony Europe Ltd, and Spectrum Bridge Incorporated. PSME frequency allocation is managed by JFMG on behalf of Ofcom. In practice, JMFG sends 'Ofcom . . . regular information on the allocation of PMSE frequencies, to enable Ofcom to provide those details to its independent database operators every three hours' (Brydon, 2013:6).

The database uses algorithms provided by Ofcom to determine white space operations to protect the first- and second-tier users, which involves ensuring that white space devices transmit at lower power in proximity to these users and more general monitoring of the system over time to respond to changes in use patterns. TVWS services in the United Kingdom thus operate under considerable regulatory constraints, at the bottom of a three-tier access system. In essence, this means that the 'speed and availability of the service cannot be guaranteed, as within this spectrum range TV broadcasters and PMSE operators take priority' (Nominet, 2018:5). Problematically, 'TVWS users share the available channels with any other users in that particular area, which may have an impact on the performance of the connection' (Nominet, 2018). The case of PMSE is interesting because, in contrast to DTT use, which tends to be very stable, 'PSME spectrum is allocated as and when required by activities in different locations. Short-term events. . . may require spectrum for as little as a few hours' (Brydon, 2013:4).

The UK experience of operating TVWS has encountered operational problems. The most difficult of these have concerned accommodation of TVWS with PMSE and wireless microphone services (author's interview, 2018). As in the United States, TVWS providers in the United Kingdom have focused for the most part on connecting rural users although this rollout has experienced problems. There is a view that for TVWS to develop its potential in the United Kingdom, aside from the technological and regulatory parameters of the system, there needs to be an attitudinal change among incumbent users of the spectrum towards a greater openness to share the space they occupy (authors' interview, 2018). The evidence of this chapter suggests that this is unlikely to occur soon.

Conclusion

Technical standards-making in international private SDOs such as the IEEE can provide potential policy solutions to pressing policy problems like the need to make more extensive and efficient use of scarce spectrum bandwidth. In the case of TVWS, this creates a context for achieving new economically lucrative and socially significant outcomes. Yet, the existence of a fractious and contested political stream—where state and regulatory steer can encounter differences of opinion strongly asserted by interests on whether and how to proceed—can impede the coupling process and the realization of an effective policy solution. The development of standards enabled DSA to exploit more intensively the use of TVWS in the United States and the United Kingdom suggests evidence of some degree of combination of the problem, policy, and political streams in the decision-making phase of TVWS policy. However, the consequences of this policy development are, at the time of writing, modest and might be regarded to some degree as precarious in terms of market development.

Changing the use of TVWS has raised classic communication policy questions around: the use of communications for public and private purposes; the value of commercial and public provision; interests of incumbent and new service providers and their users; and a broader debate about the role and position of broadcast services in an increasingly online communications environment. The chapter illustrates the potential of TVWS services in the United States and the United Kingdom, yet whilst there has been development in the UK TVWS market, there is uncertainty around whether or not it market will take off (authors' interview, 2018) and more broadly in Europe, the United Kingdom has attempted to promote TVWS spectrum sharing at the EU level and in the CEPT 'without much traction' (authors' interview, 2018). The US market too is in need of development at scale (authors' interview, 2018).

Yet, it is possible that DSA enabled TVWS services can be further developed. In emerging economies, for example, TV bands are far from full and here it is possible to use TVWS technology, even in densely populated urban areas. It is important to note that WiFi commenced at a small scale initially (authors' interview, 2018). However, to get a TVWS ecosystem in place, devices need to be more affordable and, beyond this, in the case of IoT, such services need to be utilized widely, need to be cheap, and need to demonstrate considerable capacity (authors' interview, 2018). In recent years, a particular interest has emerged in connecting rural communities in developing economies through TVWS services. Microsoft has created the Rural Airband Initiative involving partnerships with telecommunications companies to develop TVWS projects in twelve US states (Warren, 2017). Its FarmBeats project has focused on IoT underpinned data driven agriculture projects utilizing TVWS (authors' interview, 2018). The use of TVWS provides an opportunity for ICT players like this to access markets with the products and services of their core businesses.

Many of these markets suffer, on the whole, from a significant degree of network underdevelopment and disconnected users. The scope for sales of goods and services, alongside broader brand penetration, is thus large. The benefit for developing economies, which may be regarded as a significant side payment in the process, is increased levels of online connectivity. Although clearly an overwhelmingly commercial endeavour, groups like the DSA have been keen to promote the social value of TVWS, noting that its members' deployments 'are helping address key policy challenges around digital and social inclusion and are improving citizen access to education, healthcare, and government services' (Dynamic Spectrum Alliance, 2014:4). Yet, at the time of writing, as illustrated in this chapter, in those territories which have led TVWS policy change, the coupling of policy problems with solutions has proved fractious and difficult, as long and strongly held interests have played out in a political stream in which state-steered preferences have been both simultaneously strongly supported and challenged.

9

Protocols and State Surveillance

Authors have demonstrated how states utilized corporations as 'points of control' to reach public policy goals in internet governance (Farrell, 2006; Benkler, 2016). This is relevant when analysing standards-developing organizations' (SDOs) approaches to state surveillance in the context of the multiple streams (MS) framework. Chapter 4 discussed the reaction of the standards community to the Snowden revelations cumulating in development of TLS 1.3 within the Internet Engineering Task Force (IETF). This chapter discusses further ways in which state security measures have affected technical standard setting.

The following sections detail the relationship between security agencies and technology companies in SDO fora with case studies on surveillance and censorship. The chapter identifies states' 'points of control' as: restrictions in cryptography research, the interception of transport layer security (TLS) traffic by security agencies; the use of 'back-door' technologies; the development of protocols by intelligence agencies; and constraints on companies via censorship. Existing academic work concludes that 'the IETF has consistently staked out a consensus position that pushes back against technologically based indiscriminate government surveillance' (DeNardis, 2015:72).

This chapter largely confirms this position. Technology companies' ideational stance against state intervention has strengthened post-Snowden. However, the development of SDO protocols is ultimately constrained by the actions of states and the national legal structures in which companies operate. The chapter details IETF work on mitigating surveillance and Internet blocking. It then investigates the continued actions of states, with case studies on MIKEY-SAKKE (Multimedia Internet KEYing—Sakai Kasahara Key Exchange) and Huawei, with analysis of reactions by the engineering community.

Problem Definition: Security Agencies and Protocol Development

Chapter 4 described how the debate around TLS 1.3 centred on cryptography. Debates around cryptography have gone hand in hand with protocol

Global Standard-Setting in Internet Governance. Alison Harcourt, George Christou and Seamus Simpson, Oxford University Press (2020).
DOI: 10.1093/law-oso/9780198841524.001.0001

development since it began in the very early days of the Internet. In the early 1970s, the UK Government Communications Headquarters (GCHQ) and the US National Security Agency (NSA) developed algorithms to encipher public keys (Abelson et el, 2015). However, this technology remained top secret. Vint Cerf, who developed TCP/IP together with Robert Kahn, had classified access to public-key encryption at the NSA, but he was unable to use it. As he explained during a Google Hangout in 2014:

> during the mid-1970s while I was still at Stanford and working on (TCP/IP), I was also worked with the National Security Agency on the design of a secured version of the internet but we used classified crypto technology at the time and I couldn't share that with my colleagues . . . If I could start over again I would have introduced a lot more strong authentication and cryptography into the system.[1]

Only later, in 1978, did public-key encryption become commercially available with the publication of the RSA (Rivest–Shamir–Adleman) algorithm. This algorithm was used to secure Internet protocols such as TLS/Secure Sockets Layer (SSL) which enabled Secure HTTP (S-HTTP), Pretty Good Privacy (PGP)[2] and GNU Privacy Guard (GPG) to encrypt files and email.

Initially, the US Government attempted to prevent the adoption and use of RSA and PGP via a key point of control: export bans (Levy, 2001). Today, commercial and open source software containing cryptography can be exported for commercial use and is governed by the Department of Commerce.[3] However, prior to 1996, the US International Traffic in Arms Regulations (ITARS) categorized cryptography as munitions requiring licenses were required to export algorithms outside the United States. This problem for protocol development was recognized early on by the Internet Architecture Board (IAB).[4]

The first indication of its use was in 1977 when Rivest, Shamir, and Adleman were planning to present their work at an Institute of Electrical and Electronics Engineers (IEEE) meeting in New York. As Landau explains: 'they received an odd letter from a "J. A. Meyer" suggesting that the presentation of their encryption proposals would violate the International Traffic in Arms Regulation, because cryptographic systems were covered by the regulation' (2015:73).

[1] https://www.youtube.com/watch?v=17GtmwyvmWE&feature=share&t=23m1s.
[2] https://tools.ietf.org/html/rfc4880.
[3] Although the export ban was removed, encryption is also affected by patents.
[4] https://www.iab.org/documents/minutes/minutes-1993/iab-minutes-1993-07-13/.

It turned out that Meyer was an NSA employee acting alone and no further action was taken. However, it led to a push from the NSA to scrutinize academic proposals for funding to national bodies such as the National Science Foundation (NSF) and the Office of Naval Research (ONR). In 1980, with the support of the Carter Administration, the NSF made a formal statement that it would fund cryptography research without NSA scrutiny. However, later, during the Reagan Administration, 'the National Science Foundation routinely allow(ed) the N.S.A. to review any request for the funding of cryptographic research'.[5]

Then in 1991, PGP's developer Phil Zimmermann was subject to a criminal investigation by the United States Customs Service in violation of the Arms Export Control Act for 'munitions export without a license'. This is because the PGP key was larger than 40 bits. The case was eventually dropped but while it was ongoing, Zimmerman published the source code for PGP in a book with MIT Press. As publications were protected by the First Amendment, exports of the book could not be blocked (Post, 2000). The First Amendment was tested again in 1995 when a case was brought against the US government which had banned the publication of a source code for the Snuffle encryption by a student at the University of California at Berkeley. The State Department claimed that Snuffle was a 'munition' under the International Traffic in Arms Regulations (ITAR).[6] In 1999, the Ninth Circuit Court of Appeals decided that software source code was defined as speech and therefore protected by the First Amendment. A similar case was decided in *Junger v Daley* in 2000 which reiterated that software source code is protected by the First Amendment. From 1999, the United States began to loosen its restriction controls on cryptography as long as recipients were not communications companies or governments (Lanau, 2015:77). In 2016, open source encryption software source code was classified by the US Department of Commerce as 'publicly available' and no longer subject to the Export Administration Regulations.[7] However, close scrutiny of cryptography development continued.

Backlash from the engineering community against the control and use of encryption research by state security agencies at the IETF began in 1995 with the adoption of the Danvers Doctor at the 32nd IETF meeting which stipulated

[5] https://www.nytimes.com/1983/03/27/magazine/the-silent-power-of-the-nsa.html.

[6] https://casetext.com/case/bernstein-v-united-states-dept-of-justice.

[7] 81 Fed. Reg. 64656, 64668 (September 20, 2016) https://www.bis.doc.gov/index.php/policy-guidance/encryption/223-new-encryption.

that all IETF protocols should operate securely. Released in 1996, the resulting aptly named RFC 1984, reminiscent of George Orwell's dystopian novel, states:

> The IAB and IESG are therefore disturbed to note that various governments have actual or proposed policies on access to cryptographic technology that either: (a) impose restrictions by implementing export controls; and/or (b) restrict commercial and private users to weak and inadequate mechanisms such as short cryptographic keys; and/or (c) mandate that private decryption keys should be in the hands of the government or of some other third party; and/or (d) prohibit the use of cryptology entirely, or permit it only to specially authorized organizations. We believe that such policies are against the interests of consumers and the business community, are largely irrelevant to issues of military security, and provide only a marginal or illusory benefit to law enforcement agencies.

DeNardis argues that the IETF took a 'moral' stance against state surveillance (2015) even though the IETF has formally denied this.[8] The second IETF proclamation against intervention in protocol development by security agencies came out of the 'Raven Debate' in 1999. As DeNardis documents, this arose out of work on the Voice Over Internet Protocol (VoIP) within the Media Access Control (Megaco) Working Group where discussion centred on whether protocols should enable legal interception of communications in the development of IP-based telephone. Telephony interception was legal via the 1994 Communication Assistance for Law Enforcement Act (CALEA) but its extension to VoIP was unclear. The IETF posted a query on the Raven public email list asking the engineering community first, 'should the IETF develop new protocols or modify existing protocols to support mechanisms whose primary purpose is to support wiretapping or other law enforcement activities', and then 'what should the IETF's position be on informational documents that explain how to perform message or data-stream interception without protocol modifications' (IETF, 2000).

Five hundred people subscribed to the list, which provoked wider public discussion beyond the IETF and the integration of consumer interest groups into the

[8] REF2804 states: 'Why the IETF does not take a moral position: Much of the debate about wiretapping has centered around the question of whether wiretapping is morally evil, no matter who does it, necessary in any civilized society, or an effective tool for catching criminals that has been abused in the past and will be abused again. The IETF has decided not to take a position in this matter, since: There is no clear consensus around a single position in the IETF. There is no means of detecting the morality of an act "on the wire". Since the IETF deals with protocol standardization, not protocol deployment, it is not in a position to dictate that its product is only used in moral or legal ways.'

debate. As a consequence, an 'Open Letter to the Internet Engineering Task Force' was sent urging the IETF 'not to adopt new protocols or modify existing protocols to facilitate eavesdropping' signed by sixty-two representatives of companies such as AT&T and Sun Microsystems, leading academics including Lawrence Lessig, and digital rights activist organizations such as the UK Cyber-Rights and Cyber-Liberties organizations, the Center for Democracy and Technology, Privacy International, the Free Congress Foundation, the Electronic Privacy Information Center, and the Electronic Frontier Foundation.[9] At its November 1999 plenary in Washington, there was a rough consensus vote by the IETF community against enabling surveillance via protocols. RFC 2804 outlines the IETF's policy on wiretapping:

> This memo explains what the IETF thinks the question means, why its answer is 'no', and what that answer means. . . The IETF restates its strongly held belief, stated at greater length in [RFC 1984], that both commercial development of the Internet and adequate privacy for its users against illegal intrusion requires the wide availability of strong cryptographic technology. On the other hand, the IETF believes that mechanisms designed to facilitate or enable wiretapping, or methods of using other facilities for such purposes, should be openly described, so as to ensure the maximum review of the mechanisms and ensure that they adhere as closely as possible to their design constraints. The IETF believes that the publication of such mechanisms, and the publication of known weaknesses in such mechanisms, is a Good Thing.

Snowden as a Window of Opportunity

As discussed in Chapter 4, in 2013 the Snowden revelations sent shock waves through the engineering community. Insecurity in SSL and its successor TLS was known but the extent of the targeting of protocol vulnerabilities by security agencies was greatly underestimated by the IETF. RFC 7624 states that 'the attacks were . . . striking in their pervasive nature, in terms of both the volume of Internet traffic targeted and the diversity of attack techniques employed'.[10]

The Snowden documents revealed that intelligence agencies particularly the US NSA and the UK GCHQ compromised protocol security to access personal

[9] https://www.epic.org/privacy/internet/letter_to_ietf.html.
[10] https://tools.ietf.org/html/rfc7624.

data via numerous surveillance programmes. The NSA's MUSCULAR programme tapped cables between Internet service providers' (ISP) datacentres. The NSA's Keyscore accessed data from multiple points and exercises a number of techniques (Greenwald, 2013). GCHQ's Tempora programme was shown to have access to circa 1,500 cables passing within the United Kingdom (Greenwald, 2014; MacAskill et al, 2013). The *Washington Post* claimed that the NSA tapped into communication links to Yahoo and Google datacentres worldwide.[11] Google denied that it was infiltrated by the NSA.[12] However, in 2016, Cisco, Fortinet, and Juniper admitted that the NSA had been successfully targeting their firewalls for years.

The bypassing of firewalls grew into more sophisticated data capture via malware and bots. Intelligence agencies allegedly modified user cryptographic software to gain access to user content such as email, chat, and VoIP.[13] The *New York Times* claimed that the NSA 'used its influence as the world's most experienced code maker to covertly introduce weaknesses into the encryption standards followed by hardware and software developers around the world' (Perlroth, 0 Larson, and Shane, 2013). One of the leading stories resulting from the Snowden leak was that security agencies developed and promoted algorithms through standards bodies which contained kleptographic 'back doors'.[14] As RFC 7624 explains, NSA modifications 'were made to ensure that keys generated using that generator could be predicted by NSA'. It goes on to state that 'there have also been reports that the NSA paid RSA Security for a related contract with the result that the curve became the default in the RSA BSAFE product line'.

This was in reference to the Dual Elliptic Curve Deterministic Random Bit Generator (DUAL_EC_DRBG) algorithm which was published by the National Institute for Standards and Technology (NIST) in 2006. DUAL_EC_DRBG was presented as a secure algorithm to the engineering community. As Landau explains, the NSA '(i)nsert[ed] vulnerabilities into commercial encryption systems'. She goes on to explain that this 'strongly

[11] https://www.washingtonpost.com/world/national-security/nsa-infiltrates-links-to-yahoo-google-data-centers-worldwide-snowden-documents-say/2013/10/30/e51d661e-4166-11e3-8b74-d89d714ca4dd_story.html?utm_term=.9fde3fb69c6c.

[12] https://www.bbc.co.uk/news/world-us-canada-24751821.

[13] https://www.washingtonpost.com/world/national-security/nsa-infiltrates-links-to-yahoo-google-data-centers-worldwide-snowden-documents-say/2013/10/30/e51d661e-4166-11e3-8b74-d89d714ca4dd_story.html?noredirect=on&utm_term=.61ff8b69a160.

[14] See work published from the Crypto 2018 Workshop on Encryption and Surveillance https://www.lawfareblog.com/perspectives-encryption-and-surveillance.

suggested that the NSA had arranged for a cryptographic algorithm Dual EC_DRBG, into which the agency had placed a back door, to be a NIST-recommended algorithm for generating pseudorandom bits' using elliptic curve cryptography (2015:72). The standard was widely adopted by industry after the NSA allegedly paid RSA Security US$10 million to adopt the algorithm.[15] Halpin also points to back-door technologies in Schnorr and Certicom protocols (2016:3).

As Landau and others argue, this was not so surprising. This was not the first time states had attempted to insert a back door into algorithms. The NSA modified IBM's Data Encryption Standard (DES) which became a Federal Information Processing Standard (FIPS) in the 1970s (Landau, 2015:75). Later, the Clipper Chip was proposed and supported by the Clinton Government to intercept telephony communications to replace PGP.[16] Its submission by the NSA to be accredited as a FIPS also met with opposition from the engineering community.

Another leading Snowden story was the claim that technology companies had been cooperating with the NSA. A story in the *New York Times* in 2013 claimed that 'companies say they were coerced by the government into handing over their master encryption keys or building in a back door'. The *Guardian* asserted that

> some companies have been asked to hand the government the encryption keys to all customer communications, according to people familiar with the government's requests. N.S.A. documents show that the agency maintains an internal database of encryption keys for specific commercial products, called a Key Provisioning Service, which can automatically decode many messages.

It also claimed that the

> NSA's Commercial Solutions Center ... invites the makers of encryption technologies to present their products to the agency with the goal of improving American cybersecurity. But a top-secret N.S.A. document suggests that the agency's hacking division uses that same program to develop and 'leverage

[15] https://www.reuters.com/article/us-usa-security-nsa-rsa/exclusive-nsa-infiltrated-rsa-security-more-deeply-than-thought-study-idUSBREA2U0TY20140331.
[16] https://www.nytimes.com/1994/06/12/magazine/battle-of-the-clipper-chip.html?module=inline.

sensitive, cooperative relationships with specific industry partners' to insert vulnerabilities into Internet security products. (Ball, 2013).

In 2013, the German magazine *Spiegel* accused the NSA of intercepting Cisco firewalls with JETPLOW software and claimed that Cisco collaborated with the NSA to insert the back door. Cisco denied the accusation.[17] The co-operation over back doors has widely been denied by industry. Alarm was expressed by Cisco's CEO to President Obama requesting 'new standards of conduct' for the NSA.[18]

Solutions Chasing Problems

Snowden opened a window of opportunity for the IETF to address surveillance directly. In 2013, the IAB published RFC 6973 on Privacy Considerations for Internet Protocols. It defines privacy based upon legal definitions, notably in principles established by the Organisation for Economic Cooperation and Development (OECD), such as Privacy by Design, and offers guidance on integrating privacy considerations into protocol development.

This was followed in 2014 by RFC 7258 which defines pervasive monitoring as 'an attack on Internet communications that makes use of access at a large number of points in the network, or otherwise provides the attacker with access to a large amount of Internet traffic'. In 2015, the IETF outlined pervasive monitoring and how to mitigate against it in RFC 7624. The RFC distinguishes between passive and active attacks, the former comprising eavesdropping wherein packets are intercepted but no modification takes place. The latter is an attack wherein a packet is modified, removed, or added to. Internet traffic is captured via monitoring tools in Internet routers or tapping into Internet links, particularly those with high capacity, or sets of strategically located routers. RFC 7624 was followed by two information documents, the first on Technical Considerations for Internet Service Blocking and Filtering in 2016[19] and the second on the Effects of Pervasive Encryption on Operators in 2018.[20] Notably, in their analysis of the

[17] https://blogs.cisco.com/news/comment-on-der-spiegel-articles-about-nsa-tao-organization.
[18] https://www.ft.com/content/a697c292-de80-11e3-9640-00144feabdc0?ftcamp=crm/email/2014?ftcamp=crm/email/2014518/nbe/ExclusiveComment/product_a2___a3__/nbe/ExclusiveComment/product&siteedition=intl#axzz325bPWKrV.
[19] https://tools.ietf.org/html/rfc7754.
[20] https://tools.ietf.org/id/draft-mm-wg-effect-encrypt-15.html.

documents, Cath and Floridi conclude that the IETF approaches privacy as 'a technical attack that undermined trust in the network . . . not as a human rights issue' (2018:10).

Subsequently, a number of standards were identified by the IETF and flagged to the engineering community as targets for state surveillance agencies. First, the IP Flow Information Export (IPFIX) Protocol was identified as it permits administrators to amass statistics about packets passing through a network. Surveillance is able to link up the 'five-tuple': source and destination addresses, protocol type, and source and destination ports. In this way, IP addresses can be linked to specific users and their devices and which other users they interact with on the web. Reverse Domain Name System (DNS) lookup is then used to retrieve the DNS names of servers which are normally static. Although this often only provides the IP address of, for example, the home of an Internet user, it can be correlated easily by surveillance agencies as they can mandate under national laws that ISPs pass the identity of the user of the IP address for targeted surveillance.

Other targeted protocols identified by the IETF included the Session Initiation Protocol (SIP) protocol, the Post Office Protocol (POP3), Internet Message Access Protocol (IMAP), and Simple Mail Transfer Protocol (SMTP). Although these IETF standards were identified by the SDO, there are a whole host of further protocols across the Internet stack which are also susceptible to surveillance. IP addresses can be identified from mail headers. The SIP protocol enables the VOIP and instant messaging on the web. POP3 and IMAP retrieve email messages from server hosts whereas SMTP sends it.

As RFC 7624 explains, '[t]he username is directly observable if any of these protocols operate in cleartext; the username can then be directly associated with the source address'. For example, the Media Access Control (MAC) address is an IEEE 802 standard which allocates unique identifiers for devices such as mobile phones, laptops, iPads, iPods, and other devices using Ethernet, WiFi, and Bluetooth networks. MAC addresses are easily traceable as they can been 'seen' on networks and tracked by correlating users' use of services such as email, website, and gaming servers. Third-party cookies placed by advertisers can also be used for identification by surveillance agencies as they track users across different IP addresses, particularly if a user has already been identified by one IP address. The determined efforts of the IETF to deal with potential compromise of IETF protocols was followed by an increase in privacy and security work within SDOs as a whole as will be covered in Chapter 10.

The HTTP Status Code 451

States exercise points of control in other ways. ISPs and browsers often receive government and state requests to take down content on the web. In response to such requests, the IETF developed the HTTP status code 451 'unavailable for legal reasons', which was adopted in December 2015. Although the initiative for the code came from a participant engineer, pressure for adoption came from the wider Internet community. This pressure reflected wide public pressure for transparency, consumer, and client privacy and consent online.

Formerly, when websites were taken down or blocked, they would show the 401 error code ('without authorisation'), 403 ('Forbidden'), or 404 ('Page Not Found'). However, the HTTP status code 451 produces a separate error code which shows when a website has been blocked by ISPs due to a government or court request or legal order. Ray Bradbury wrote *Fahrenheit 451* in 1953. In this futuristic novel, books are banned and burned by 'firemen', but the main protagonist, a fireman Guy Montag, disillusioned with his work, leaves to join a secret society committed to the preservation of books and to re-educate society. For this reason, the choice of the number 451 was in every sense a political statement by the engineering community, signalling clear dissatisfaction with government censorship.

In 2014, Tim Bray, an Android developer working at Google, submitted a draft to the IETF working group.[21] Bray's idea for the 451 code came from a blog post[22] by Terence Eden at the UK Open Rights Group when he pointed out that there was a lack of an error message for censorship when he received a 403 error when trying to access the Pirate Bay which is blocked in the United Kingdom. The idea for 451 was not new and had been raised by other authors prior to this time.[23]

There was initial reluctance by the Chair of the IETF HTTP Working Group, Mark Nottingham, and the Applications Area Director, Barry Leiba, when first approached by Bray. Leiba did not recognize a strong technical case as the proposal did not meet the criteria for new status codes.[24] Mark Nottingham pointed out that 400 status codes are limited. As Nottingham explained in a blog post: '[i]nitially, I and some others pushed back. HTTP status codes are a constrained name space; once we use everything from 400 to 499, for example,

[21] https://tools.ietf.org/id/draft-tbray-http-legally-restricted-status-00.html.

[22] https://shkspr.mobi/blog/2012/06/there-is-no-http-code-for-censorship-but-perhaps-there-should-be/.

[23] http://www.qwghlm.co.uk/2008/12/09/there-is-no-http-code-for-censorship/.

[24] https://httpwg.org/specs/rfc7231.html#considerations.for.new.status.codes.

we're out of luck'.[25] Leiba also took into consideration concerns within the IETF community that 451 could be viewed as a 'warrant canary'. A warrant canary is a colloquial term for a web status update wherein an ISP alerts users to the fact that it has received a letter from a state security agency to take down web content without approval from a court. Public outing of such requests can be met with criminal penalties, for example, in the United States under Section 18 U.S.C. §2709(c) of the USA Patriot Act.

The 451 proposal was considered by the IETF community in November 2015. Support for the initiative was garnered externally with evidence for demand and use. The code was heavily discussed on IETF mailing lists and in meetings. Most feedback came via hosting organizations such as Github, Google, and Firefox extensions. Digital rights groups such as Lumen and Article 19 argued that the utility of having a machine-readable marker could aid them in identifying censored websites on the Internet. As stated by Article 19, 'sometimes a simple paragraph of computer code can help media developers fight online censorship'.[26] Another argument was that 451 would alleviate other forms of blocking. As Kuchta explains in his discussion of 451, site blocking

> may involve blocking the URL (Uniform Resource Locator) entered in the browser (e.g. http://www.website.com/), also including links to specific content (e.g. http://www.website.com/subsite/file.pdf)—a type of block that is very precise but also costly and difficult to implement. Another method is to block the domain name (e.g. website.com) or the associated Internet Protocol (IP) address (e.g. 123.45.678.90), but this prevents access to all content at the given domain name or IP address, legal or illegal (known as 'overblocking').[27]

Most 451 discussions referred to the blocking of free speech. Another form is the blocking of copyrighted and trademarked material. The Open Rights Group approach related most to this form of the protocol's use as the group has consistently campaigned against blocking orders in the United Kingdom regardless of origin as it views them as setting a potential precedence for the blocking of restrict free speech.[28] This includes copyright restriction based upon territory.

[25] https://www.mnot.net/blog/2015/12/18/451.

[26] https://www.article19.org/resources/status-code-451-an-internet-governance-standard-to-expose-censorship/.

[27] http://www.codozasady.pl/en/error-451-site-unavailable-for-legal-reasons/.

[28] https://www.openrightsgroup.org/press/releases/trade-mark-case-could-set-a-precedent-for-website-blocking.

On balance, the Area Director felt that the IETF community was in favour of the 451 status code and the specification was agreed in December 2015 in RFC 7725. Discussion within the IETF revolves around whether a request for taking down a website should come from a court order in order to justify use of the status code and how to distinguish between a government order and a court order. The webpage software also needs to be configured to utilize 451 and it is complicated to implement in HTTPs. Not all web servers are prepared to implement the code for this reason. Since its adoption, the code has been used extensively to signal when a website cannot be accessible for copyright reasons such as by newspaper websites when accessed from outside the territory licensed for copyright.

The Internet Research Task Force's (IRTF) Human Rights Protocol Considerations Research Group (HRPC), in which many digital rights groups participate, took on the responsibility of monitoring implementation of the code. In 2017, the group reported on 451 implementation in Australia, China, the EU, Japan, Kazakhstan, Norway, Russia, the Philippines, Thailand, the United Arab Emirates, and the United States in 2017.[29] The 2017, the Prague IETF worked on RFC 7725 that specifies the new HTTP status code 451 and different implementations in the Hackathon. The Hackathon developed a plug-in to help users to report use of 451 to the IETF, a 'crawling tool' to identify legally withheld web content,[30] a repository, and developed a plugin for Wordpress authors for geoblocking purposes.

The Snowden revelations also drew attention to security agency work within the United Kingdom. The upgrade to 4G-enabled increased surveillance in communications standards. Mobile phone companies introduced a call encryption standard (called Secure Voice). The standard was agreed at the Third Generation Partnership Project standardization group (3GPP) and advanced by the United Kingdom (specifically GCHQ's Communications-Electronics Security Group (CESG)). It transpired that the standard included a protocol (called MIKEY-SAKKE) which GCHQ allegedly can switch off.

Development of MIKEY-SAKKE began in 2004 through a security protocol for real-time multimedia applications by engineers at Ericsson. In 2008, the International Telecommunication Union-Radio communications sector (ITU-R) agreed on the International Mobile Telecommunications Advanced (IMT-Advanced) specification (required for the 4G standard). This necessitated

[29] https://tools.ietf.org/id/draft-451-imp-report-00.html.

[30] The tool was developed by the NetBlocks Open Source Internet observatory project https://github.com/ntblk/block-crawler.

the need at the UK level to upgrade the existing TETRA (Terrestrial Trunked Radio) standard to the 4G Public Safety system, the Long Term Evolution (LTE) and (Public Safety-LTE) standards for emergency services (fire, police and ambulance). TETRA was a European standard developed by ETSI for emergency, rail, transport, and military services and recognized by the European Radiocommunications Committee.

However, CESG at GCHQ chose instead to adapt the IETF's MIKEY-SAKKE for security use rather than the standard agreed within ETSI. The MIKEY-SAKKE protocol was developed by Michael Groves at CESG and published as an open source IETF protocol in 2012 for use with UK emergency services (without by that time participation of Ericsson). The standard was adopted by the Secure Chorus technology group for interoperability in secure voice communications by the UK-based groups Armour Communications, BT, CESG, Cryptify, Cyber Y, Finmeccanica, Metaswitch, Samsung, SQR Systems, and Vodafone. The prime intended use was for security in congested advanced networks with end-to-end encryption, with public safety a core goal.

In 2016, a security researcher at University College London alleged that GCHQ had introduced a back door.[31] The claim was that the ability for network providers to replicate private keys and decrypt phone calls could be subject to warrants to interception or subject to hacking. The protocol also enabled network providers to collect metadata and the identity of callers. Finally, it offered organizations the option to monitor their employees via recording encrypted communications then decrypting them. GCHQ responded that MIKEY-SAKKE was not intended for surveillance.

Criticism by digital rights groups of GCHQ continues. In 2019 an open letter[32] was signed by forty-seven digital rights groups including Big Brother Watch, the Center for Democracy and Technology, and Privacy International, and companies including Apple, Google, Microsoft, and WhatsApp to GCHQ against their 'ghost protocol' proposed by UK National Cyber Security Centre's technical director Ian Levy and GCHQ's chief codebreaker Crispin Robinson in November 2018. The proposal was for messages to be copied and sent on to third recipients which would then would possess a 'key' to unencrypt the messages. The letter was flagged in the media and Levy replied that the idea was only ' "hypothetical" and intended as a "starting point for discussion" '[33].

[31] https://www.bbc.co.uk/news/technology-35372545.

[32] https://newamericadotorg.s3.amazonaws.com/documents/Coalition_Letter_to_GCHQ_on_Ghost_Proposal_-_May_22_2019.pdf.

[33] https://www.bbc.co.uk/news/technology-48458710.

The Case of Huawei

Concern about back doors to enable state espionage is, of course, not isolated to the United States and the United Kingdom. In May 2019, President Trump issued Executive Order 13873 Securing the Information and Communications Technology and Services Supply Chain. The Order prohibits:

> any acquisition, importation, transfer, installation, dealing in, or use of any information and communications technology or service (transaction) by any person, or with respect to any property, subject to the jurisdiction of the United States, where the transaction involves any property in which any foreign country or a national thereof has any interest (including through an interest in a contract for the provision of the technology or service)

The order was clearly aimed at Huawei, a major Chinese technology company involved in the development of broadband and mobile phone networks globally as, after the order was issued, the Department of Commerce added Huawei to the 'Entity List' which 'identifies foreign parties that are prohibited from receiving some or all items subject to the EAR unless the exporter secures a license'.[34] This means, among other things, that Qualcomm cannot provide chips to Huawei. Google has also severed Huawei's Android access to application and security updates and is restricting Android use in Huawei Mate X and P30 Pro. It is thought that Microsoft may block Windows 10 updates for Huawei laptops.

The main suspicion is that Huawei has included a back door in its technology to allow for Chinese state-sponsored espionage. For this reason, development of Huawei's technology for use in US and EU markets is closely monitored. Formal US accusations against Huawei are manifold. First, Huawei employees were accused of stealing a T-Mobile robot arm from a lab in Seattle in order to copy the technology. Huawei's parent company in China allegedly offered incentives to employees to steal trade secrets.[35] Second, Huawei's chief financial officer and the daughter of the company's founder, Meng Wanzhou, was arrested in Canada in December 2018 at request of the United States under an extradition treaty. Meng was charged with bank fraud and trading with Iran against US sanctions policy.[36] The main concern is that China is reselling

[34] https://www.bis.doc.gov/index.php/documents/regulation-docs/691-supplement-no-4-to-part-744-entity-list/file.

[35] https://arstechnica.com/tech-policy/2019/01/us-indicts-huawei-for-stealing-t-mobile-robot-selling-us-tech-to-iran/.

[36] https://www.justice.gov/opa/press-release/file/1125021/download.

US technology to Iran. In a similar case, the Chinese smartphone company ZTE was fined US$1.19 billion for exporting US technology to Iran and North Korea in 2017.

In January 2019, Daniel Coats, US Director of National Intelligence, told the Senate Select Committee on Intelligence that 'the potential [exists] for Chinese intelligence and security services to use Chinese information technology firms as routine and systemic espionage platforms against the United States and allies'.[37] One month later, in February 2019, US Vice President Mike Pence stated '[w]e must protect our critical telecom infrastructure, and America is calling on all our security partners to be vigilant and to reject any enterprise that would compromise the integrity of our communications technology or our national security system'. This was followed by the Executive Order in May.

The United States has long put pressure on states to ban Huawei from operating in communications markets due to potential back-door technologies which would permit the Chinese government to spy or conduct cyber-attacks. The main states under pressure are members of the 'Five Eyes' transnational alliance, under which the NSA operates together with GCHQ, the Australian Signals Directorate (ASD), the Canadian Communications Security Establishment (CSEC), and the New Zealand's Government Communications Security Bureau (GCSB) (Bigo, 2019). GCSB barred Huawei from its National Broadband Network in 2012 and both Australia and New Zealand have banned Huawei from participating in 5G.

Within the European Union, France banned Huawei from its core network provision. However, Huawei's hardware is in use by BT's EE and Vodafone's core networks in the United Kingdom, BT being the first company within Europe to purchase Huawei equipment. However, in 2019, BT removed Huawei equipment from parts of its 3G and 4G networks and Vodafone suspended its purchase of Huawei equipment for 5G networks in Europe. At the time of writing (2019), the UK Government is considering blocking Huawei from 5G networks under US pressure. Germany decided not to ban Huawei, however, from its networks in 2019.

The United Kingdom has been monitoring Huawei's presence in the UK's telecommunications networks since 2014 when it established its Huawei HCSEC oversight board, chaired by the head of GCHQ's National Cyber Security Centre, which produces annual reports to the UK's National Security Adviser. In its 2019 report, the Huawei Cyber Security Evaluation Centre (HCSEC) oversight board expressed grave concerns over Huawei's supply of

[37] https://www.dni.gov/files/ODNI/documents/2019-ATA-SFR---SSCI.pdf.

equipment to the UK's 5G networks.[38] It did not find evidence of a back door as such but pointed to weaknesses which could permit surveillance. The issue is highly politically sensitive. The UK defence secretary, Gavin Williamson, was sacked in May 2019 after he allegedly revealed sensitive information about the UK's plans for Huawei during a National Security Council meeting. The discussions allegedly concerned whether Huawei should be able to supply 'non-core' 5G mobile phone equipment in the United Kingdom.

Conclusion

Problem definition in this chapter was framed by state agencies on the one hand, whose prime remit is surveillance and the censoring of illegal content, and by engineers working in SDOs on the other, whose core goal is security and encryption. A window of opportunity to address security flaws in the IETF work on privacy was opened by the Snowden revelations. This enabled the adoption of (often pre-existing) solutions to mitigate surveillance via protocol development within the IETF. However, legal demands continue to facilitate state surveillance on the Internet. DeNardis suggests that the IETF's backlash to collaboration with security agencies was in some respect a choice between transparent and non-transparent interception as 'individual companies, such as router manufacturers (could still) be required, under CALEA in the United States or by various laws in other countries, to build lawful intercept capacity into products' (2015:79).

Due to the backlash by technology companies to government surveillance, greater transparency has been injected into security surveillance requests. As a result of a campaign by the Reform Government Surveillance group, supported by key technology companies (notably Microsoft, Facebook, Google, and Yahoo), the US Department of Justice changed its terms on public disclosure of data requests by surveillance agencies via the FISA court in an effort to foster transparency in January 2014.[39] The companies are permitted to publish the number of requests received and whether they relate to content or non-content after a six-month period. Companies such as Microsoft[40] and Google[41] now publish this data publicly. This has enabled data analyses by

[38] https://assets.publishing.service.gov.uk/government/uploads/system/uploads/attachment_data/file/790270/HCSEC_OversightBoardReport-2019.pdfHUAWEI.
[39] https://www.justice.gov/iso/opa/resources/366201412716018407143.pdf.
[40] https://www.microsoft.com/en-us/corporate-responsibility/fisa.
[41] https://transparencyreport.google.com/user-data/us-national-security?hl=en.

blogs on forums such as Github[42] and consumer rights organizations[43] such as the American Civil Liberties Union. This was followed by a statement in a UN Special Report to the United Nations Human Rights Council in 2018 recommending that '[s]tates should publish detailed transparency reports on all content-related requests issued to intermediaries and involve genuine public input in all regulatory considerations'.[44]

The demand for information from technology companies is not expected to lessen. In January 2018, the US Senate voted to renew Section 702 of FISA. Section 702 of the 2008 Foreign Intelligence Surveillance Act (FISA) Amendments Act was the statutory basis for legalizing government surveillance programmes such as PRISM. This permitted US security agencies to access data from ISPs such as Comcast and technology companies such as Google through the FISA court. Since 2012, the Act has included 'Procedures for Targeting Certain Persons Outside the United States Other Than United States Persons'. In 2019, a letter was sent to Facebook from from Attorney General William P. Barr, United Kingdom Home Secretary Priti Patel, Australia's Minister for Home Affairs Peter Dutton, and Acting Homeland Security Secretary Kevin McAleenan requesting Facebook CEO, Mark Zuckerberg, "not proceed with its end-to-end encryption plan".[45]

The evolution of IoT has exacerbated the tension between intelligence services and technology companies. In the United States, it was stated that social media might be used as an additional platform for monitoring, surveillance, and tracking. In 2016, James Clapper, Director of National Intelligence, informed the US Senate Armed Services Committee that in the future, 'intelligence services might use the [internet of things] for identification, surveillance, monitoring, location tracking, and targeting for recruitment, or to gain access to networks or user credentials'.

This has prompted further backlash to government intervention from US technology companies. A letter sent to the House of Representatives in November 2017 requested greater respect for consumer privacy from a number of companies including Amazon, Google, Twitter, and IBM. In February 2018, at the World Economic Forum, Ginni Rometty, Chairman, President, and Chief Executive Officer of the IBM Corporation, stated that 'IBM . . . will not

[42] https://github.blog/2019-01-23-2018-transparency-report/.
[43] https://www.aclu.org/sites/default/files/field_document/cjeu_schrems_report_final_august_30_2018.pdf.
[44] https://freedex.org/a-human-rights-approach-to-platform-content-regulation/.
[45] https://www.justice.gov/opa/pr/attorney-general-barr-signs-letter-facebook-us-uk-and-australian-leaders-regarding-use-end.

provide client data to any government's surveillance programme. We support our clients' choices on where their data is stored. And we ensure their data doesn't go anywhere without their consent.' Certainly, this indicates that the engineering community is ideationally opposed to any involvement with government surveillance. The next chapter will look at current developments by technology companies and coordination within SDOs to tighten consumer privacy.

10
Engineers and the Public Interest

As Chapter 5 demonstrated, recent years have seen a high degree of spillover of policy domains formally defined at the level of the state to standards-developing organizations (SDOs). This represents a shift in the political architecture away from national legislation and international treaty-based agreements to technical standard agreements in fora such as the World Wide Web Consortium (W3C). Although the literature on activists and their relationship with private sector global governance is extensive, little attention has been paid to their participation within SDOs.

To further address this gap, this chapter examines the participation of digital rights and consumer groups in SDO work on unsanctioned tracking within the W3C and IETF protocol development. The chapter demonstrates how public interest initiatives can come to fruition within SDO fora. It shows how SDO work is flanked and supported externally by academics, digital rights groups, and wider civil society organizations. However, as protocol adoption is self-regulatory, action is limited to best practice and monitoring and discourse is shaped to a great extent by external legal environments.

Problem Definition: Unsanctioned Tracking

Most public interest initiatives within the W3C stem from unsanctioned tracking. Public revelations about personal privacy intrusions by social media and online platforms are becoming increasingly frequent, not least the Cambridge Analytica Facebook scandal in 2018. However, the extent of personal data sharing between companies has only recently come to light. In 2019, the Federal Trade Commission brought a platform and user data practices case against Facebook, accusing it of sharing user data with over 150 companies including Apple, Amazon, Microsoft, Netflix, Sony, and Spotify. This was preceded by a *New York Times* investigation which claimed that contact information and even private messages were shared with Netflix and Spotify without

Global Standard-Setting in Internet Governance. Alison Harcourt, George Christou and Seamus Simpson, Oxford University Press (2020).
DOI: 10.1093/law-oso/9780198841524.001.0001

user consent.[1] In reaction to these developments, the UK parliament seized a cache of internal company documents from Facebook. The documents, which were made public, showed how difficult Facebook made it for consumers to realize that there had been a privacy update enabling the collection of personal data from Android mobile phone users' calls and texts.[2]

Digital rights groups are drawing increasing attention to problems of unauthorized online behavioural online advertising and real time bidding. In September 2018, the Open Rights Group, a representative from the browser company Brave together with an academic from University College London filed complaints against online behavioural advertising with the UK's Information Commission Office, the Irish Data Protection Commission, and the French Commission nationale de l'informatique et des libertés (CNiL).[3] The submissions mainly targeted Google. This was followed by Privacy International filing complaints with against seven data brokers (Acxiom, Oracle), ad-tech companies (Criteo, Quantcast, Tapad), and credit referencing agencies (Equifax, Experian) with the same data protection offices. Then, in 2019, the digital rights group, the Panoptykon Foundation, filed complaints against Google and the Internet Architecture Board (IAB) Europe to the Polish Data Protection Authority.

However, other forms of tracking are less well known. Users are tracked remotely based upon their device activity. Devices such as computers, cameras, and mobile phones can be identified by indicators such as screen size, software versions, and installed fonts (Eckersley, 2010; Gunes, Juarez, Nikiforakis, Diaz, Gürses, et al., 2013:1130). JavaScript code identifies audio settings, battery status, installed plugins, location and time settings, and screen resolution fonts. This is called 'browser fingerprinting'. Commercial software can 'fingerprint' web browsers and analyse use without users' knowledge even when these users are aware enough to block browsing cookies and/or use virtual private networks (VPNs) (Doty, 2019).

The Electronic Frontier Foundation (EFF) documented the effectiveness of fingerprinting via the Panopticlick project in 2010 (Eckersley, 2010). The project used voluntary participants through wide advertising, particularly via distribution on social media websites such as Twitter and Facebook. It fingerprinted browsers via markers such as fonts, time zones, and 'accept'

[1] https://www.nytimes.com/interactive/2018/06/03/technology/facebook-device-partners-users-friends-data.html?module=inline.

[2] https://www.parliament.uk/documents/commons-committees/culture-media-and-sport/Note-by-Chair-and-selected-documents-ordered-from-Six4Three.pdf.

[3] https://brave.com/adtech-data-breach-complaint.

headers. Digg and Reddit found that, of the 470,161 browser users who voluntarily visited the website http://panopticlick.eff.org, 83.6% could be uniquely identified (Eckersley, 2010). This was followed by a number of studies, notably the DIVERSIFY project funded by the European Union at INRIA which ran the amiunique.org fingerprinting site and the WebTap project at Princeton University. The studies raised significant interest and concern from the engineering community which prompted a number of initiatives within SDOs.

Unsanctioned tracking occurs across devices and platforms. As defined by Peacock, it is the 'unseen and unauthorised extraction, storage, analysing, selling, buying, and auctioning of personal online data appropriated by one or more remote online corporate actors' (2014:1). Internet users are not only unaware of tracking, but are unable to opt out. Personal data include online blogs, photographs, instant messaging and texts, Tweets, the content of emails, videos and the metadata attached to them (Peacock, 2014:2). As Ryan notes, 'even if bid request traffic is secure . . . there are no technical measures in place to adequately protect the data.'[4] The data is used by big data companies for a wide range of purposes, from behavioural advertising to assessing credit ratings and insurance risk.

Anti-fingerprinting techniques have been developed by the Tor Uplift project[5] and have since been implemented by Mozilla,[6] Brave,[7] and Apple.[8] Apple, Google, Mozilla, and Tor have dedicated pages to developers on how to minimize fingerprinting.[9] Mitigation comes in the form or differential privacy (DP) which is implemented via two models: centralized differential privacy (CDP) and local differential privacy (LDP) (Chowdry, et al., 2019). Clearly, the pressure to adopt these measures comes from the legal environment. As Chowdhury and colleagues point out, Google uses CDP 'to detect changes in browser properties of its userbase as it does not want the legal burden of storing highly sensitive browser fingerprints in the clear on its server' (Chowdhury, Wang, He, Machanavajjhala, and Somesh, 2019:2). Google has also developed Encode, Shuffle, Analyze (ESA) to mitigate the monitoring of computer users' software activities and protect user privacy.[10] Chrome and Apple browsers utilize LDP.

[4] https://brave.com/Behavioural-advertising-and-personal-data.pdf.

[5] https://www.ghacks.net/2016/07/04/tor-privacy-settings-coming-to-firefox/.

[6] https://www.ghacks.net/2019/03/06/firefox-67-with-anti-fingerprinting-technique-letterboxing/.

[7] https://github.com/brave/browser-laptop/wiki/Fingerprinting-Protection-Mode.

[8] https://gizmodo.com/apple-declares-war-on-browser-fingerprinting-the-sneak-1826549108.

[9] Apple: https://trac.webkit.org/wiki/Fingerprinting; Google: https://sites.google.com/a/chromium.org/dev/Home/chromium-security/client-identification-mechanisms#TOC-Fingerprinting-prevention-and-detection-challenges; Mozilla: https://wiki.mozilla.org/Fingerprinting; Tor: https://2019.www.torproject.org/projects/torbrowser/design/#fingerprinting-linkability.

[10] https://ai.google/research/pubs/pub46411.

The Technical Architecture Group of the W3C has documented 'Unsanctioned Tracking: Tracking without User Control' identifying browser fingerprinting, SuperCookies, and header enrichment as key techniques.[11] Browser fingerprinting is defined as 'the capability of a site to identify or re-identify a visiting user, user agent or device via configuration settings or other observable characteristics' (Doty, 2019). As described by the W3C's Privacy Interest Group (PING), browser fingerprinting can be used for tracking users across the web through browsing activity across devices. A fingerprint is established which is unique to a single user. It can then be correlated with email addresses, surnames, or other identifiers such as government issued IDs to build up a profile of Internet use without user consent.

PING classifies fingerprinting into passive, active, and 'cookie-like'. Passive fingerprinting normally involves cookies which identify the browser and operating system. Active fingerprinting is when code is run on a user's computer to deliberately document the browser, user and/or device including tracking performance characteristics. Cookie-like fingerprinting enables re-identification of a user by storing information on the user's device normally with super, zombie, or evercookies which are permanent.[12] The clearing of cookies or switching devices does not mitigate this problem. As Gunes and colleagues (Gunes et al., 2013) explain a 'device fingerprint may be collected and compared to a database of known devices. Using such techniques, known devices can be matched, and previously unknown devices can be added to the database.' IETF Internet standards are designed to accommodate Hypertext Transfer Protocol (HTTP) cookies as documented first in Request for Comments (RFC) 2109 and most recently in RFC 6265. Privacy problems associated with cookies and fingerprinting were recognized from early on in the IETF but there was little willingness to take action. Peacock attributes this to 'the impact of (the IETF's) accommodating institutional framework' (Peacock, 2014:6). In this respect, standardization is viewed as neutral and not designed to support privacy invasion as such.

Most of the work to mitigate online tracking has been done within the W3C, mainly within PING and the Tracking Protection Working Group. As detailed in Chapter 7, the Tracking Protection Working Group led to the Do Not Track (DNT) specification. This group was closed in 2019 after the DNT specification was formulated and compliance is monitored by the Tracking Compliance

[11] https://w3ctag.github.io/unsanctioned-tracking/.

[12] HTTP cookies for HTML were developed by Lou Montulli from Netscape in 1994 to mitigate 'memory loss' but have evolved substantially since this time: https://samy.pl/evercookie/.

and Scope Working Group.[13] PING continues work on browser fingerprinting post-DNT operating as a W3C Interest Group. The Group recommends best practice in technical specifications to mitigate tracking via fingerprinting. Most recommendations advise that developers minimize the risk by limiting access or making fingerprinting more difficult or exposing it so that users are aware that they are being tracked. Specific advice relates to limiting access to devices for application use. This includes limiting access to microphones and cameras after users have granted access but also limiting access to device metadata. Standardization or common implementation has also been recommended so that there is consistent practice across the engineering world.

In 2015, Lukasz and colleagues drew the attention of the W3C's Device and Sensors Working Group to tracking through device batteries (Lukasz, Gunes, Castelluccia, and Diaz, 2016). Using data provided by Firefox and Linux, they demonstrated how a 'fingerprintable surface' could be used to track users and their location. Shortly after this, in May 2016, Keith Chen, head of economic research at Uber, admitted on National Public Radio's 'Hidden Brain' programme that Uber had been using the data to charge more for taxi hire when users' batteries were dying. Englehardt and Narayanan expanded this work to show that user identify can be exposed via manipulating WebRTC, audio, and battery status application programme interfaces (APIs) (Englehardt and Narayanan, 2016). For example, third-party trackers can use WebRTC to expose local IP addresses on Ethernet and WiFi and track users when they are using real-time communication applications such as Skype or Facetime. Halpin notes 'the IETF/W3C WebRTC reveals the IP address of its user even if a VPN is being used' (2016:3).

This exposure led to the withdrawal of companies from APIs from source codes in browsers. It began with withdrawal of the Battery API (a proposed Recommendation by the W3C) by Mozilla, Firefox, and Apple Safari in 2016. The move was welcomed by privacy advocates and digital rights groups.[14] As stated by Lukasz in a blog post, this marked 'an unprecedented event in the Web's history. Web browser vendors are deciding to reduce browser's functionalities—remove features—due to privacy risk potential. We went a long way since 2000s, when privacy wasn't treated that seriously.' The profiling of the Battery API by the W3C has drawn greater attention by developers to privacy issues.

13 https://www.w3.org/TR/tracking-compliance/.

14 https://groups.google.com/forum/#!topic/mozilla.dev.platform/5U8NHoUY-1k/discussion.

In 2017, the W3C developed the WebCrypto API together with the Web Hypertext Application Technology Working Group (WHATWG) within the Web Cryptography Working Group, which provides 'basic cryptographic operations in web applications, such as hashing, signature generation and verification, and encryption and decryption.'[15] The standard began in a community group based on Mozilla's DomCrypt, cryptography API. As Yates and Murphy highlight: 'the committee's work was of critical importance to online commerce, and it touched on the deeply contentious issues of security across the internet and the World Wide Web' (2019:269).

Discussions on WebCrypto revolved around whether the standard should operate at a 'high level' (more complex) or 'be easier and safer for developers to use' and whether or not algorithms should be mandatory for use with the spec (Yates and Murphy, 2019:274). A more 'high-level' API was chosen based upon preferences on mailing discussion lists. However, as Mozilla warns, 'the Web Crypto API provides a number of low-level cryptographic primitives. It's very easy to misuse them, and the pitfalls involved can be very subtle.'[16] The W3C Recommendation was flanked by an earlier 2014 RFC 7230 at the IETF which mitigates the identification of user devices. The W3C Web Cryptography Working Group closed after its Recommendation was agreed, and is maintained by the Web Security Interest Group. The Web Security Interest Group scrutinizes and reviews specifications throughout the W3C. The Group is also working on standardization for watermarking and mitigating fingerprinting. An interviewee stated that this work aimed 'to balance functionality of the web with user privacy'.

The Web Authentication Working Group has been developing a number of solutions to user privacy. It became apparent that tracking was becoming more pervasive with the introduction of Hypertext Markup Language (HTML) 5.2 which increased the use of APIs. HTTPs and WiFi encrypt traffic but, due to the increase in phishing and mass data breaches, the W3C is currently looking at using devices as replacement for passwords. Standards are needed for devices to communicate with computers. The W3C is working together with members of the Fast IDentity Online (FIDO) Alliance[17] to produce authentication standards. FIDO produced the Client to Authenticator Protocol (CTAP) protocol which determines how a computer talks a device. The Web Authentication Working Group built a consortium with five main browser

[15] https://www.w3.org/TR/WebCryptoAPI/.
[16] https://developer.mozilla.org/en-US/docs/Web/API/Web_Crypto_API.
[17] https://fidoalliance.org/.

developers working on devices: Google, Mozilla, Microsoft, Ubico, and Qualcomm which are working on chips in mobile communications. Paypal is also working on a device. The issue surfaced within W3C community groups which included digital rights group participants and members of the Web Security Interest Group. The National Institute of Standards and Technology (NIST), the US Federal Reserve, the UK Government, and New Zealand send representatives as observers. In this respect, web browsers will no longer have to store data which are not useful anywhere else. They only utilizes the data that the user provides to it such as age verification.

In 2019, the W3C published WebAuthn as a W3C Recommendation. WebAuthn was developed by FIDO to standardize applications and services using public-key cryptography. The first version has been implemented in Windows 10, Android, and Chrome, Firefox, Edge, and Safari Web browsers. Development is underway in 2019 for a second version. In interviewees, some W3C members stated that it would be easier to promote privacy if there were legal support mechanisms in place supporting the work within standard-setting fora. There is scope for this in the proposed ePrivacy Regulation (ePR) of the European Union.[18] The proposed Regulation states that '[i]nformation related to the end user's device may also be collected remotely for the purpose of identification and tracking, using techniques such as the so-called "device fingerprinting", often without the knowledge of the end-user, and may seriously intrude upon the privacy of these end-users'. However, implementation may be difficult as the Article 29 Working Party's Opinion 01/2017 on the proposal points out:

> any interference with the terminal equipment, including the use of its processing capabilities, requires the consent of the end user . . . The EC has now helpfully confirmed that 'device fingerprinting' falls under this provision. In addition, the Working Party welcomes that the failure of a third party to abide by the preferences expressed in an individual's browser settings are enforceable as described in recital 22. This is helpful for those situations where a third party (e.g. an ad network) does not respect these settings. However, this should also be laid down in a relevant provision of the Proposed Regulation.

In 2018, Tim Berners-Lee announced the SOcial LInked Data (SOLID) programme.[19] It is an academic programme but 'relies as much as possible on

[18] https://ec.europa.eu/digital-single-market/en/news/proposal-regulation-privacy-and-electronic-communications.

[19] https://solid.mit.edu/.

existing W3C standards and protocols'. The programme introduces Pods which store personal data for use across devices. This allows them to 'decouple' their data from the Internet. This would support proposals within the EU's privacy Directive and allow for data portability as stipulated as a right in Article 20 of the GDPR.

The W3C's Verifiable Claims Working Group has also been working on a Verifiable Credential which it describes as 'an object a Holder can present to a Verifier showing assertions made by an Issuer'. It would contain metadata, an optional identifier, and a cryptographic proof for identity. This has attracted companies from the ledger industry. For example, the Sovrin Foundation, a non-profit organization, which handles blockchain in a public/private way, as follows. The cryptography is based on that of IBM. Distribution is based upon a 'zero-knowledge technology'. It allows users to view cryptographic objects selectively and present certain information such as age or whether the use is a member of a club. In this respect, it is founded on self-identity based on user choice and discards the role of inspector. The negative side of the scheme is, however, that if users share a key, others can access everything that user has. Other companies offer similar solutions on a profit model basis such as Everynym and Drummond Reed. A number of W3C conferences have been run on the issue such as Rebooting Web of Trust and Internet Identity workshops. The W3C has been seeking open standardization via an API with a proposal in the Credentials Community Group since 2015.

The IETF and Consumer Privacy

Most of the IETF work to mitigate pervasive monitoring which also serves to protect consumer privacy was described in Chapter 9. This section outlines further work which is specifically related to consumer privacy. Pressure to integrate consumer privacy into protocol development came from a wide number of digital rights groups but particularly from the Centre for Democracy and Technology which identified privacy gaps at the IETF from 2003 on.[20] Since then, the IETF has run an increasing number of working groups which focus on consumer privacy. From 2010, the IAB, W3C, and IETF worked jointly via the Privacy Program following a workshop organized with the W3C, the Internet Society, and the Computer Science and Artificial Intelligence Laboratory (CSAIL). The RFC was authored by Alissa Cooper who was then working for

[20] https://www.iab.org/wp-content/IAB-uploads/2011/03/alissa_cooper.pdf.

the rights group, the Centre for Democracy and Technology (CTD), and later became Chair of the IETF in 2017. The RFC stated privacy risks resulted from 'information leakage, difficulties in distinguishing first parties from third parties, complications arising from system dependencies, and the lack of transparency and user awareness of privacy risks and trade-offs'. The Privacy Program was replaced by the Privacy and Security Program in May 2014 which authored RFC 7696 as described in Chapter 9.[21]

After 2003, a number of working groups were established to identify privacy risks in protocol development. The Geographic Location/Privacy (GEOPRIV) group began work in 2004. It mainly worked on securing location data during voice, video, and messaging applications (in the use of SIP) (Rosenberg et al, 2002). The group closed in 2014 and was replaced with the SIPREC group which worked on a protocol developed to suppress metadata in the recording of online voice and sessions with the use of the Session Initiation Protocol (SIP), Session Description Protocol (SDP), and Real-Time Transport Protocol (RTP) 'out of concern for privacy or perceived lack of need for it to be included in the recording' (RFC 7866).

The Application-Layer Traffic Optimization (ALTO) and the Congestion Exposure working groups arose out of a 2008 workshop investing congestion issues in Peer-to-Peer (P2P) traffic (RFC 5594). ALTO was set up to protect users from the sharing of content preferences and network topologies in the sharing of data. Since this time, it has developed a large number of protocols aimed at protecting consumer privacy.

Another important IETF initiative was Certificate Transparency (CT) which is an experimental IETF open standard and open source framework for monitoring and auditing digital certificates. An engineer at Google put forth a proposal within the IETF for making CA operations transparent which was supported by the Internet Society as well as the ACLU and EFF. The work began in 2013 with RFC 6962 and was finalized in 2018.[22] It was a response to a number of certificate scandals, particularly Diginotar, OpenSSL Heartbleed, and Apple's Goto Fail. Certificates for TLS/ SSL come into play between a website and an end user's browser which contain the server's public cryptographic key. As Arnbak and colleagues explain (Arnbak, Asghari, van Eeten, and van Eijk, 2014):

> First, a trust relationship (a handshake) is established between a website and an end user's browser. This is done with the help of a Transport Layer Security/

[21] https://www.iab.org/activities/programs/privacy-and-security-program/.

[22] https://tools.ietf.org/html/draft-ietf-trans-rfc6962-bis-30.

Secure Sockets Layer (TLS/SSL) certificate containing basic information for authentication purposes. If the Web browser trusts the certificate and the issuing CA, then this authentication handshake succeeds. Second, successful authentication leads to a TLS/SSL-encrypted channel between the website and browser, called a tunnel. Thus, the handshake authentication serves as the stepping stone for the confidentiality and integrity that HTTPS seeks to deliver. If the handshake succeeds, then the browser informs the user by, for example, depicting a padlock or a green address bar. If the TLS/SSL certificate or the issuing CA cannot be trusted, then the browser will show a security warning to the end user.

Websites acquire certificates from certificate authorities which hold the host name, certificate owner, issuer, period of validity, and a public key. It is easy for any company to set up a certificate authority (CA) as there is no market regulation. In 2011, the CA provider DigiNotar, which had strong presence in the Dutch market particularly for government e-services, issued around 500 digital certificates including a number for the Dutch Government. This included CAs for all major websites globally including Google, Mozilla, Facebook, and Yahoo which were then used by the Iranian Government for spying purposes. The leak became apparent only after some months so the hack was not known about for a long time. Users were exposed for around six weeks. Interviews revealed that companies were never able to find the full list of certificates issued. There were no records because it was sufficiently hacked. As a result, a number of difficulties were experienced in the Netherlands for some weeks. For example, ships could not clear customs electronically meaning that shipping containers' admission to the country were delayed. There were also problems with electricity provision and the sale of cars.

If companies had been aware of which certificates were responsible they could have targeted the faulty ones and left the others alone but they were not. This is when the need for the standard first became apparent. The CT protocol agreed within the IETF aims to weaken the risk of an attack by making all TLS certificates publicly available via certificate logs. These logs are in turn auditable.

As Cheval and colleagues explain, 'the main idea behind the protocol is not to prevent a certificate authority to misbehave but to be able to detect when (it) did misbehave' (Cheval, Delaune, and Ryan, 2014:88). Since this time, because CA certificates remain as yet unregulated, the industry has engaged in self-regulation. Google and the Mozilla Found maintains a list of trustworthy CAs for Chrome[23] and Firefox[24] respectively. Apple and Microsoft refer to Chrome's list for Windows, macOS, and iOS as they use Chrominium. Symantec sold its certificate authority to DigiCert when Google threatened to stop recognizing them because Google felt that Symantec was insecure. The move to TLS 1.3 should mitigate certificate problems.

The CA case shows that much of the mitigation work done within the IETF is reactive. The Domain-based Message Authentication, Reporting & Conformance (DMARC) specification is another example of this. It is an email authentication protocol for instructing servers whether to bounce or receive incoming mail if it fails authentication under Sender Policy Framework (SPF) and DomainKeys Identified Mail (DKIM). SPF and DKIM prevents senders from sending email on another user's behalf.[25] The FTC found that 86% of technology companies use SPF (Federal Trade Commission, 2017).

A number of companies[26] joined together in 2010 to form DMARC with the goal of 'combating fraudulent email at Internet-scale'.[27] The goal was two-fold: 'to enable senders to publish easily discoverable policies on unauthenticated email (and) . . . receivers to provide authentication reporting to senders so that they can improve and monitor their authentication infrastructure'. They aimed to develop an interoperable specification which would then be submitted as a formalized standard. Some members of the group had already been loosely working together on solutions up until this time. The DMARC specification was agreed between consortium members in 2012.

Meanwhile, in 2014, Yahoo and AOL changed their DMARC policies to 'reject'. This meant that if emails did not come from their own respective (Yahoo or AOL) servers, the email would not get through to the recipient. This was implemented to prevent phishing and spoofing. As a result, an IETF working group was set up in 2014 to address DMARC and interoperability. However, the RFC was only published as an informational document in 2015.[28]

The Web Authorization (OAuth) protocol agreed in 2017 in the IETF's Open Web Authentication group[29] permits users to grant an API access to

[23] http://www.chromium.org/Home/chromium-security/root-ca-policy.

[24] https://wiki.mozilla.org/CA/Included_Certificates.

[25] https://docs.microsoft.com/en-us/office365/securitycompliance/use-dmarc-to-validate-email.

[26] Receivers: AOL, Comcast, Gmail, Hotmail, Netease, Yahoo! Mail Senders: American Greetings, Bank of America, Facebook, Fidelity, JPMorgan Chase & Co., LinkedIn, PayPal Intermediaries & Vendors: Agari, Cloudmark, ReturnPath, and the Trusted Domain Project Industry. The groups also liaised with the Technology Policy Division of the Financial Services Roundtable (BITS), Messaging, Malware, and Mobile Anti-Abuse Working Group (M3AAWG), and the Online Trust Alliance (OTA).

[27] https://dmarc.org/about/history/.

[28] https://tools.ietf.org/html/rfc7489.

[29] https://datatracker.ietf.org/wg/oauth/about/.

the user's protected resources with the necessity to reveal their identity. It was promoted within the IETF by Twitter and Google from 2010 (RFC 5849). The 2017 protocol has been adopted by Amazon, Facebook, Instagram, LinkedIn, Microsoft, Netflix, and Paypal, amongst others. Interviews revealed that security companies were less satisfied with OAuth as they did not want SML but JavaScript Object Notation (JSON). However, developers were dissatisfied with JSON as it did not have the semantic richness needed. This led to the OpenID protocol being created and used in a separate consortium external to the W3C.

In 2016, the Privacy Enhancements and Assessments Research Group (PEARG) was established by the IRTF which aims at 'reviewing privacy enhancing technologies for network protocols and distributed systems'.[30] It is not yet involved in protocol development but acts as a forum for discussion. In particular, it integrates open source and academic communities into the IETF in the formation of informational and experimental RFCs and to document research on ongoing privacy work. PEARG aims to engage with external groups such as the EFF, Tor Project, EFF, and OTF and forums such as Usable Privacy and Security (SOUPS), the W3C, the Privacy Interest Group, and Privacy Enhancing Technologies Symposium (PETS). However, to date, it has relied upon a small number of researchers, one of whom is supported via Article 19's Internet of Rights Fellowship Program. Amongst other work, PEARG worked on anti-fingerprinting, statistical inference, multi-party computation (MPC) for privacy, preserving reputation systems, and Google's Encode, Shuffle, Analyze (ESA) in 2019.

Digital Rights Activism

Digital rights groups have not been highly active within the IETF. A small number of groups can be seen at IETF meetings notably the ACLU, Article 19, the Center for Democracy and Technology, and the EFF. These are supported by wider groups in campaigns and notably on discussion fora. Most pressure is exerted through supporting one position or another within SDO fora as documented in the TLS 1.3 and Encrypted Media Extensions cases. In this respect, digital rights and consumer groups have acted as policy entrepreneurs mobilizing support and framing topics (Princen, 2007).

These groups have, however, taken issue with a number of groups operating within the IETF, most notably Cisco which allegedly cooperated with China

[30] https://datatracker.ietf.org/doc/charter-irtf-pearg/.

in the development of China's Golden Shield project (otherwise known as the 'the Great Firewall').[31] A number of law suits were filed against Cisco[32] contending that 'that the Golden Shield, as designed and implemented by Cisco, enabled the Chinese security services to compile a database of information regarding the victims' individual circumstances (patterns of life, family composition, political tendencies, work history, financial information, etc.) to inform its practice of *zhuanhua*—the forced conversion of dissidents through torture and mistreatment.[33]

Digital rights groups have targeted the IETF engineering community and companies involved in IETF work more directly from 2016 on. Article 19 and the Asociación por los Derechos Civiles (ADC) showed films at the 2016 Berlin 95 IETF in Buenos Aires: the Net of Rights and Computers and; the Remarkable Story of the ENIAC Programmers. At the 2016 Berlin 96 IETF, a representative from the Centre for Democracy and Technology wrote a 'censorship draft'[34] for the Security Area Group to demonstrate to engineers how their protocols have been used by governments for censorship purposes.

In 2016, a group of digital rights groups[35] alleged that Verisign was threatening Internet users' rights in the development of the Extensible Provisioning Protocol (EPP).[36] The group claimed that Verisign was developing a Registry Verification Framework within the IETF to comply with China's draft law on Internet Domain Name Management Rules which presented a threat to the right to privacy and freedom of expression online.[37] This related to the implementation of real name policies for domain name system registration in China. In their statement, the groups cited the rights to privacy within the UN General Assembly, the UN Human Rights Council, and the UN Special Rapporteur on Freedom of Expression.[38] IETF work ceased once flagged by digital rights groups.

[31] https://cs.stanford.edu/people/eroberts/cs201/projects/2010-11/FreedomOfInformationChina/background/index.html.

[32] http://www.sanjoseinside.com/wp-content/uploads/2017/05/Doe-v-Cisco-Amended-Complaint.pdf.

[33] http://cyberlaw.stanford.edu/blog/2015/03/china%E2%80%99s-golden-shield-cisco-systems-complicit.

[34] https://tools.ietf.org/html/draft-hall-censorship-tech-03.

[35] Article 19, the Asociación por los Derechos Civiles (ADC), Association for Progressive Communications (APC), Bytes for all, the Centre for Internet and Society (CIS), Derechos Digitales, Digital Defenders Partnership (Hivos), Internet Democracy Project, Reporters Without Borders, SonTusDatos, TEDIC, and the Web Foundation.

[36] https://tools.ietf.org/html/draft-ietf-regext-verificationcode-00.

[37] https://www.article19.org/resources/corporate-actors-must-not-facilitate-human-rights-violations-through-new-chinese-rules/.

[38] https://www.yumpu.com/en/document/read/55758267/irtf-rg-human-rights-protocol-considerations-hrpc.

This case demonstrates that, although national legal efforts might assist SDO work on personal privacy, disparities between different jurisdictions can hamper efforts. Even divergences between the US and EU can create problems for SDOs. An example of this is the discrepancy between the United States and the European Union over WHOIS requirements which run counter to the obligations of the 2016 EU General Data Protection Regulation (GDPR). WHOIS was established in 1982 and specified that each Advanced Research Projects Agency Network (ARPANET) host had a US mailing address, zip code, and telephone number (Bygrave, Schiavetta, Thunem, Lange, and Phillips, 2009:158). Today, WHOIS requires that registries provide domain name owner and technical contact details to be published publicly with an online registry. In 2018, a legal case was brought against WHOIS in Germany by EPAG domain registrar, which attempted to sell domain names without providing personal data details. EPAG argued that these requirements would incur fines under the GDPR and the German court agreed. As a result, the International Corporation for Assigned Names and Numbers (ICANN) published an Advisory Statement, Temporary Specification for generic top level domain gTLD Registration Data, in May 2018 which stated that registries will still need to collect all information but the online WHOIS database will only provide 'thin data' for European requests.[39]

Conclusion

Grugel and Iusmen (2013) argue that there has been a rise in rights activism, but it is difficult to get such issues on to the international agenda. This chapter has shown that the participation of digital rights activism can bring about ideational change in internet standard setting governing SDO participant behaviour. Changes are particularly evidenced with progress within the W3C and IETF on mitigating unsanctioned tracking. This SDO work is clearly based on wider underlying societal norms concerning the right to transparency and privacy. Similar to the DRM case discussed in Chapter 3, the protection of personal privacy is yet another case of spillover to SDO fora of a policy domain which has historically belonged to the state, .

[39] https://www.icann.org/en/system/files/files/advisory-statement-gtld-registration-data-specs-17may18-en.pdf.

11

The Internet of Things

A Policy Window for Standard Essential Patents

Apple iPhones do not use the standardized micro-Universal Serial Bus (USB) port but specialized ports for their proprietary Lightning connectors.[1] However, in October 2018, John Ternus, Apple's vice-president of hardware engineering, announced the switch to USB-C for the new generation of iPads from 2019. Apple iPhone is expected to follow suit. This marks the end to a long-raging battle over port standardization for personal computers within the Institute of Electrical and Electronics Engineers (IEEE). The change came about largely due to a push from the European Commission for industry to agree upon a universally proprietary free standard[2]

Software patents have long been a bone of contention within SDOs. Allegedly, 'patent trolls' have been the subject of wider public discussion such as the One-click-buy patent (Amazon), high-speed memory chips (Rambus), JPEG (Forgent), and modified statistical coding of digital signals (General Electric). For this reason, states have been increasingly involved in steering intellectual property rights (IPR) policies within standards-developing organization (SDO) fora. The European Commission in particular is identified as a policy entrepreneur which identified the problem of SDO's granting preference for the licensing of standard essential patents (SEPs) to end product makers. Civil society too has had its role to play from the stance of open source solutions to interoperability albeit with less and varying degrees of success across the fora. This chapter focuses on how the European Union and the United States have driven change to IPR policies within the World Wide Web Consortium (W3C), the Internet Engineering Task Force (IETF), and the IEEE. State pressure has counterbalanced historic corporate pressure for preserving patent protection.

[1] The micro-USB is a standard developed within the IEEE.

[2] https://ec.europa.eu/info/law/better-regulation/initiatives/ares-2018-6427186/feedback/F18119_en?p_id=342389,

Global Standard Setting in Internet Governance. Alison Harcourt, George Christou and Seamus Simpson, Oxford University Press (2020). © Alison Harcourt, George Christou, & Seamus Simpson.
DOI: 10.1093/law-oso/9780198841524.001.0001

The Internet of Things (IoT) is identified within the chapter as a systemic shock mechanism leading to the strengthening of state actors within SDOs and a window of opportunity for change. Once this window was opened, it enabled decision-makers within SDO fora to push for a more open IPR approach. It sheds light on how states can shape actor behaviour within SDOs and under what conditions they can bring about policy change. State steer is identified as a macro mechanism (Gerring, 2010) as it relates to institutional and external pressure. The key mechanism for policy change described in this chapter is therefore institutional. It should be noted, however, that state action in this case has not reflected any formal legal convergence between the US and EU but demonstrates soft power initiatives from entrepreneurial figures within US and EU administrations at key points in time.

SDOs are working within the context of wider government architectures. Historically, discrepancies between EU and US intellectual property laws have created institutional ambiguity affecting preferences within SDOs (Weston and Kretschmer, 2012). The United States has recognized the patentability of software since 1988[3] and software can be patented with the US Patent and Trademark Office. However, EU countries ban patents on software, formulae, and algorithms according to the 1973 European Patent Convention Article 52(c), which excludes 'programs for computers' from patentability (Drahos, 2010). EU-based companies concerned about a level playing field with US groups lobbied for a European Patent Directive (Guadamuz, 2006). The Directive on the Patentability of Computer Implemented Innovations (the 'Software Patent Directive') was rejected by the European Parliament in 2005 (Haunss and Kohlmorgen, 2010). When the EU Unitary Patent Package was agreed in 2012, the rule against patenting software remained in place. At the same time, US patents are recognized in Europe under the 1994 Agreement on Trade-Related Aspects of Intellectual Property Rights (TRIPS). This incompatibility between different legal settings has prompted forum shopping in the registration of patents. Most patents in the high-tech industry are registered in the United States and indeed the Information, Communications and Technology (ICT) sector represents the most numerous number of filings globally (Table 11.1 and Figure 11.2).

[3] This was first recognized in 1981 in *Diamond v Diehr* where mathematical equations could be patented: https://supreme.justia.com/cases/federal/us/450/175/case.html, and then in 1988, in *State Street Bank & Trust Co. v Singature Financial Group Inc*, that software could be patented.

Table 11.1 Intellectual property rights disclosures in the IETF

2018	199
2017	133
2016	132
2015	182
2014	163
2013	297

Source: https://datatracker.ietf.org/ipr/

Governance Architecture

IPR policies within SDO fora differ in nature. The W3C stipulates that standards are patent free;[4] the IETF promotes open standards; whereas the Organization for the Advancement of Structured Information Standards (OASIS) promotes open standards but also adopts patented royalty-free standards. Forum shopping takes place between different SDOs based on IPR preferences, and IPR often decides where patent pools form. Booz Allen Hamilton, BEA Systems, Intel, and Oracle have sought interoperability solutions with OASIS due to patent issues.[5] Once an SDO has been identified by a group of companies, they tend to remain with the forum for long-term agreements on interoperability. Often Special Interest Groups (SIGs) form within SDOs to work on a specific standard, as documented in Chapter 6. As Drezner notes, 'private actors with constrained exit options have a strong incentive to invest in assets specific to longstanding domestic legal and regulatory structures' (Drezner, 2008:5).

Standard setting is a soft form of governance that relies on decision-making processes in the absence of centralized control. Most SDO fora in the field of Internet governance are dominated by large corporations, with minimal to no formal representation from states and little input from end-users or third-sector groups. However, states do possess tools to influence SDO decision-making. SDOs operate within the context of wider international

[4] In 1996, the W3C agreed the Portable Network Graphics (PNG) standard which offered a patent-free replacement for the Graphics Interchange Format (GIF) (Alverstand and Lie, 2009:140). Disagreements over the licensing GIF between the software patent holders, Unisys, and CompuServe persisted within the W3C until 2004 when the patents expired. This was despite long campaigning by open source and digital rights groups against patent use, e.g. 'Burn All GiFs'.

[5] These companies originally sought interoperability via the Web Services Interoperability (WS-I) which was set up by IBM and Microsoft in 2002. In 2017, WS-I was incorporated into OASIS http://www.ws-i.org/.

legal systems which shape their decision-making. Decision-making in the W3C, the IETF, and the IEEE are permissible under international and national agreements and agencies. In the United States, these are chiefly the 2004 US Standards Development Organization Advancement Act (SDOAA), the 2012 Leahy–Smith America Invents Act, the American National Standards Institute (ANSI),[6] and the WTO's Technical Barriers to Trade (TBT) Committee.[7]

In Europe, these include the European Commission's Communication on Intellectual Property Rights and Standardization of 1992 and the 2010 Horizontal Cooperation Agreements. The EU's Horizontal Guidelines state 'the assessment of whether fees charged for access to IPR in the standard-setting context are unfair or unreasonable should be based on whether the fees bear a reasonable relationship to the economic value of the IPR'.[8]

The EU's 2012 Regulation on European standardization maintains that standards are defined as 'voluntary technical specifications for products, production process, or services'. Patent use can also be directly challenged under competition law and in the courts.[9] In recent years, the US Supreme Court and European Court of Justice have become involved in outlining IPR requirements within SDOs. This was flanked with recent proclamations and statements from the International Telecommunications Union (ITU), the European Commission, and the US Department of Justice. These tools act as sticks with which to steer SDO fora decision-making.

Participation in SDOs is voluntary; decisions take place on a self-regulatory basis between participants or members. However, patents are highly contentious. They are proprietary in nature and may constrict potential software development. Patents last for twenty years if registered in the United States or the European Union.[10] SEPs are crucial components of standards development. These are standards which manufacturers cannot develop and sell technologies without utilizing. This gives companies that own SEPs significant market power.

For this reason, SDO IPR policies exist to prevent incentivizing behaviour, royalty stacking, patent ambush, and hold-ups (Régibeau, De Coninck, and

[6] For a list of SDOs accredited by the American National Standards Institute, see: https://www.ansi.org/standards_activities/domestic_programs/accreditation_as_developer/asd.

[7] The TBT has a code of good practice in Annex 3 https://www.wto.org/english/docs_e/legal_e/17-tbt_e.htm#annexIII. A Directory lists all SDOs which have signed up to the Code https://www.iso.org/files/live/sites/isoorg/files/archive/pdf/en/wto-tbt-scd.pdf.

[8] Commission Communication on Guidelines on the applicability of Article 101 of the Treaty on the Functioning of the European Union to horizontal co-operation agreements (text with European Economic Area relevance) (2011/C 11/01).

[9] A prime example of this is the 2000 Sun/ETSI case wherein Sun was sanctioned for declaring an SEP post adoption of the ETSI GSM 03.19 standard and the 2010 Rambus DRAM-chips cases (COMP/38.636). See http://ec.europa.eu/competition/publications/cpn/2011_1_1_en.pdf for discussion.

[10] https://www.law.cornell.edu/uscode/text/35/154.

Zenger, 2011; Lundqvist 2014:164–5). Royalty stacking is a situation where a standard contains many SEPs, owners of which each charge a royalty for use. Patent ambushes occur where SEP owners withhold ownership until after a standard is agreed. In this case, the royalty rates 'reflect a monopoly position' (Lundqvist, 2014:165). Hold-ups are situations where SEP owners demand excessive royalties only after users have made substantial cost investments, producing 'lock-in'. As Kühn and colleagues explain: 'hold-ups occur when the SEP owner approaches firms practicing the standard—after those firms have invested in developing their products . . . with an onerous licensing demand' (Kühn, Morton, and Shelanski, 2013:3).

In the 2012 *Microsoft Corp v Motorola* case, the US Court of Appeals noted that 'there are at least 92 entities that own 802.11 (SEPs). If each of these 92 entities sought royals . . . the aggregate royalty to implement the 802.11 standard, which is only one feature of the Xbox product, would exceed the total product price' (Contreras, 2016:858). In the 2018 *Tech. Holdings, Ltd. v Telefonaktiebolaget LM Ericsson* case, 'Ericsson estimated that an essentiality assessment of its portfolio of 2,600 SEP families pertaining to ETSI's 2G, 3G and 4G standards would have cost approximately $26 million' (Contreras, 2017:8).

Each SDO has its own IPR policy. As stated earlier, W3C and IETF standards must be royalty free. OASIS promotes open standards but also adopts patented royalty-free standards. Royalty-free patent policies do not mean that SDO Working Groups are not affected by IPR. Standards often incorporate and build on existing standards which may have pre-existing IPR commitments. Hence, SDO policies relate not only to the final product—a royalty-free standard—but the disclosure, licensing arrangements, and use/prohibition of an existing IPR of standards that the new standard incorporates. Most SDOs require fair, reasonable, and non-discriminatory (FRAND) commitments, whereby a company licenses a patent on FRAND terms, in order to be utilized.

IETF

Today's IETF's IPR policy is outlined under its RFC 8179 introduced in May 2017. IPR is declared in the IPR data tracker.[11] Contributions to Working

[11] https://datatracker.ietf.org/ipr/.

Groups (WGs), Birds of a Feather (BOFs), the Internet Engineering Steering Group, the Internet Architecture Board, IETF mailing list, or Internet drafts function (Requests for Comments (RFCs)) must adhere to copyright, license, and disclosure rules. These are designed, according to RFC 8179, 'to ensure that IETF working groups and participants have as much information as possible about any IPR constraints on a technical proposal as early as possible in the development process'. Every meeting is prefaced by a 'Note Well' which reiterates the IPR policy. Although transparent, Halpin notes that the IETF note well is not as transparent as the W3C (2016:2). If contributors are aware of any patent held by their own employer or anyone participating, this needs to be disclosed. As the IETF is made up of individual participants rather than company representatives, they are not expected to know about all IPRs owned by their employers. However, often Working Groups are arranged around IPR issues under the form of 'super groups'. This means that late disclosure is rare in the IETF compared to other SDOs. However, certain sectors tend to have higher IPR disclosures, for example, the television and music sector. The IETF's music WG showed thirty-five IPR disclosures between 2005 and 2015.

As Contreras describes, the IETF's original 1992 policy (RFC 1310) adopted the language of ANSI's patent policy emphasizing FRAND (2016:870). It evolved to RF or FRAND licensing requirements in 1994 (RFC 1602). Simcoe maintains that IPR was not a serious issue within the IETF until 1995 when Motorola disclosed patents covering Compression and Encryption Control Protocols (2009:271). Lundqvist claims that the 1993 NCRPA 'marked a turning point when SSOs had to take on this (IPR) responsibility' (Lundqvist, 2014:165). Clearly a number of factors triggered a rethink and in 1996 the IETF's policy was revised with RFC 2026. This stipulated that participants disclose patents but they did not require a specific form of licensing. Importantly, the 1993 NCRPA 'create(d) a safe haven for SDOPs to discuss IP issues and, even, royalty rates' (Lundqvist, 2014:165). As Simcoe documents, this resulted in a rise in disclosures given the requirement (see Table 11.1). RFC 2026 required anyone who made a contribution to a standard to disclose any patents held by the contributor or his/her employer or of any third-party patents which were 'reasonably personally known' to the participant. However, participants were discouraged from discussing patent policies on mailing lists but could do so on an informal basis, preferably in hallways outside of Working Groups (Contreras, 2016:873).

Contreras documents thirty-seven legal disputes over patents involving IETF standards between 2003 and 2015 (2016:882–895). This includes protocols central to the web such as Domain Name System (DNS), Hypertext Transfer Protocol (HTTP), Internet Protocol (IP), Session Initiation Protocol (SIP), Simple Mail Transfer Protocol (SMTP), Secure Sockets Layer (SSL), Transport Layer Security (TLS), and Resource ReSerVation Protocol (RSVP). Given the rise in legal disputes, the IETF's policy was subsequently revised in 2017 after substantial internal discussion. Significantly, the practice of informal discussion of patents outside Working Groups was reversed. Any contribution to the IETF triggers the requirement to disclose patents including discussion on mailing lists. Disclosure can be required based on oral contributions as well as mailing list discussion or 'in any other way acting in order to influence the outcome of a discussion relating to the IETF Standards Process'. Bradner and Contreras assume 'thus, silently rolling your eyes or giving a thumb's down during a technical presentation could subject you to the IETF's disclosure requirements' (Bradner and Contreras, 2017). The 2017 change also mandates detail around the process for updating IPR disclosures. In addition it outlines areas where RF is preferred, for example, in security areas. Finally, and most importantly, voluntary statements about terms on which licences cover patents are now regarded as binding and irrevocable.

IEEE

As Bekkers notes, the IEEE was the only SDO (the other being VITA[12]) which 'received the non-binding approval of the US Department of Justice to provide for ex ante disclosure of patent licensing terms by its members prior to voting on approval of a standard' (2013:26). The IEEE adopted a new patent policy in 2015. The policy stipulated a calculation of FRAND royalty costs and a ban against injunctions opposing SEPs. Unlike the IETF and W3C, the IEEE permits calculations on royalty standards but agreements stipulate IPR use. IPR disclosure of essential patents is required during two stages of the IEEE decision-making process. Under Stage I, patents must be disclosed in the Project Authorization Request (PAR) which is approved by the New Standards Committee (NesCom) and the Standards Board (IEEE-SA). Stage II consists of the formation of a Working Group (WG) and call for participation at which stage disclosure of essential patents is again requested. Stage II progresses to the production of drafts which are posted publicly for a period of sixty days

[12] VITA was began in 1982 as the Virtual Machine Environment (VME)bus Manufacturers Group. It changed its name to VMEbus International Trade Association (VITA) in 1984.

and then balloted when agreement by 75% of WG members is reached and after all comments have been responded to. Prior to the new policy, the IEEE's policy was based upon an ex ante disclosure of SEPs and FRAND committees (Kanevskaia and Zingales, 2016:20). Like the IETF, the IEEE formerly prohibited discussion of patents within working groups.[13]

The most salient patent dispute for the IEEE was led by Microsoft, Intel, and Apple for IEEE 802.11 wireless networking standards. It began in 2010 when Motorola increased licensing costs for IEEE 802.11 patents underpinning WiFi for mobile phones and the ITU-T H.264 patent for encoding (used in Microsoft's Xbox). Microsoft filed a case against Motorola in 2010. Apple and Intel sided with Microsoft and filed briefs in the case. Qualcomm and Nokia supported Motorola. In 2012, Google bought Motorola Mobility. Most authors argue that this was in order to acquire Motorola's patent portfolio including 802.11 and H.264. In 2013, *Microsoft Corp. v Motorola* was decided by a district court in Seattle. The judge specified that RAND should be used as a licensing term for an SEP portfolio, and set a specific royalty rate and range. It was the first time a US court stipulated the use of RAND in a standard-setting case.

5G is the most important area for the IoT as it is expected to link up a wide range of 'things' (from farming equipment to ships) to electronic devices. According to Strategy Analytics, the Qualcomm, Nokia, and Ericsson companies will account for over 90% of the future 5G smartphone market (see Figure 11.2).[14] Whereas 3G and 4G mainly dealt with patents for the mobile phone sector, IoT brings together a wide range of sectors from utilities, transport, and healthcare to manufacturing. This represents the need for a wider pooling of SEPs. There has been a rapid increase in 5G patents in recent years (Figures 11.1 and 11.2). Major SEP owners are AlcatelLucent, Ericsson, Nokia, InterDigital, Qualcomm (Mallinson, 2015).

The most important IoT area for the IEEE is low-power wide-area network (LPWAN). LPWAN is a new technology to be used with 5G. It can operate in licensed and unlicensed frequency bands and link up a large number of devices across the network. The advantage is that it allows devices to connect in ways that do not use a lot of power. The WG is working on enabling IPv6 connectivity with the LPWAN technologies: SIGFOX (a French company), the LoRa alliance, WI-SUN (a global consortium of companies), and NB-IOT.

[13] http://standards.ieee.org/develop/policies/antitrust.pdf.

[14] https://www.strategyanalytics.com/access-services/devices/mobile-phones/device-technologies/about-emerging-devices.

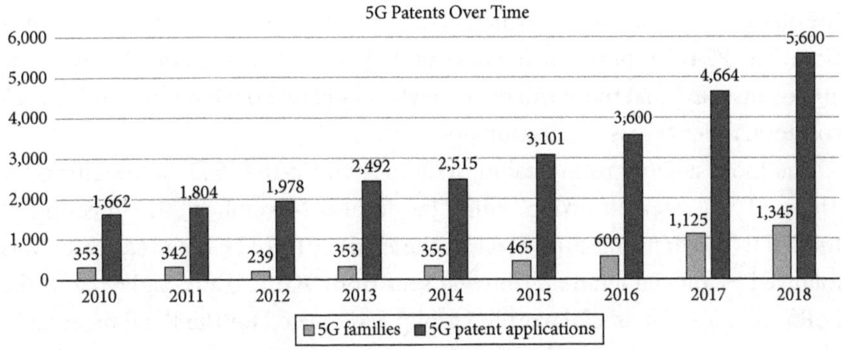

Figure 11.1 5G patents over time.

Source: IPlytics GmbH[a]

[a] https://www.iplytics.com/report/sep-auto-industry-5g/ (last accessed 19 April 2019). IPlytics—Tim Pohlman. More up-to-date statistics are available from https://www.iplytics.com with information on standard essential patent owners which include more granted and non-granted SEPs.

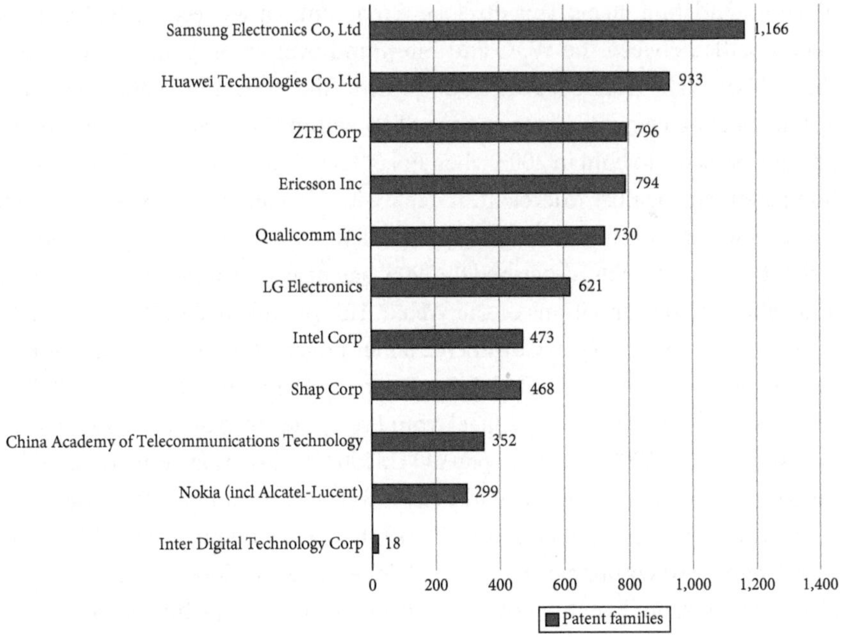

Figure 11.2 5G SEP owners.

3GPP is planning submission of NB-IoT and eMTC (LTE-M) as a use case to the IMT-2020 at the ITU.

The IEEE is also important for IoT as it established the 802.15.4 standard for short-range, low-power radios. The new version of the standard introduced the 802.15.4g physical layer, which allows for larger packets transmission over IPv6. The P2413 group which works on IoT was established in 2015 with the aim of ensuring that there will be no new architecture or standards and to seek commonalities across application domains.

The most salient group dealing with IoT within the IETF is the Thing-to-Thing (T2T) research group which began at a November 2015 meeting of the IETF. The first BoF took place at the April 2016 IETF meeting which was followed by official letters of interest sent from 3GPP, IEEE, SigFox, and the LoRa Alliance. An official working group was created for the IETF meeting in November 2016 in Seoul. The WG also serves as contact point for other SDOs.

W3C

Within the W3C, IPR is not formally disclosed at the outset but at the 'Recommendation' stage. This disclosure requirement was established in 1999 after a battle between the W3C and Intermind over an SEP contained within the P3P standard (Russell, 2011:166). Prior to this time, the W3C relied on a 'gentleman's agreement' not to pursue SEPs within the W3C. A second high-profile case came to light in 2003 when Eolas Technologies won a US$520.6 million patent suit against Microsoft. It dealt with US patent US5838906[15] which was granted to Eolas in 1998 for a protocol which embedded applications in web browsers. The W3C described the '906' patent as 'one of the most important and widely used foundations of the Web'.[16] This provoked Tim Berners-Lee to write to Under Secretary of Commerce James Rogan requesting invalidation of the patent.[17] In 2004, the United States Patent and Trademark Office (USPTO) invalidated the patent. Under appeal from Eolas, the 906 patent was re-enacted in 2005. Then in 2009, a second patent (US2003154261) was granted based on the same technology which would permit websites to embed applications via plug-ins and AJAX (JavaScript and XML). Ebolas subsequently filed lawsuits with twenty-two companies including Adobe Systems, Google, Yahoo, Apple, eBay, and Amazon. Eventually, in 2012, the two patents were invalidated by the US Court of Appeals for the Federal Circuit.

[15] Distributed hypermedia method for automatically invoking external application providing interaction and display of embedded objects within a hypermedia document. See: https://patents.google.com/patent/US5838906A/en.

[16] https://www.w3.org/2004/10/patents-standards-innovation.html.

[17] https://www.w3.org/2003/10/27-rogan.html.

Other key patents for the W3C include the Microsoft Style Sheet Patent and SunXLink Patents. Microsoft's Style Sheet US5860073 patent affects development of the W3C's Cascading Style Sheet (CSS). Microsoft agreed

> that any W3C member or other party will be able to obtain a license under Microsoft's US Patent No. 5,860,073 to implement and use the technology described in W3C Recommendations for Cascading Style Sheets (CSS) and the EXtensible Style Language (XSL) for the purpose of supporting such standards, on a royalty-free basis. One condition of this license shall be the party's agreement to not assert patent rights against Microsoft and other companies for their implementation of those standards. Microsoft expressly reserves all other rights it may have in the material and subject matter of this contribution.[18]

Sun's XLink US5659729 patent affects development of the XLink specification. Sun also developed a license for royalty-free implementation of the standard.

The W3C introduced a royalty-free policy for W3C standards in 2004 after lengthy consultation with members.[19] Initially, an RF/FRAND policy was proposed but resistance, particularly from open source groups, brought about its revision (Russell, 2011:159). However, like the IETF, the W3C puts together 'super groups' based on IPR issues and occasionally, patent advisory groups (PAGs) are set up within the W3C to assess IPR cases. Since 2004, the W3C has maintained a public list of patent disclosures.[20] A list of active WGs is listed at the bottom of the webpage. Under section 6 of the W3C Patent Policy, members must disclose patent information.[21] The Patent and Standards Interest Group (PSIG) was established as a forum to discuss the W3C patent policy as well as wider issues dealing with 'patents, licenses and Web standards'.[22] The W3C's Web of Things, established in 2017, aims are to find 'potential for using scripting languages like JavaScript, data encodings such as JSON and EXI, formats for data and metadata, including Linked Data, and protocols such as HTTP and WebSockets, etc.'

[18] https://www.w3.org/Style/CSS/Disclosures.en.html.
[19] https://www.w3.org/Consortium/Patent-Policy-20040205/.
[20] https://www.w3.org/2004/01/pp-impl/83482/status.
[21] https://www.w3.org/Consortium/Patent-Policy-20170801/#sec-Disclosure.
[22] https://www.w3.org/2004/pp/psig/charter-2019.html.

WHATWG

As discussed in Chapter 5, the W3C HTML standards are often preceded by the Web Hypertext Application Technology Working Group (WHATWG) which has formulated its own Intellectual Property Rights Policy.[23] Like the W3C, WHATWG has a 'Patent-Licensing Obligation' which includes a 'Contribution Licensing Obligation' and 'Review Draft Licensing Obligations'. The policy was developed within WHATWG by Apple, Google, Microsoft, and Mozilla in 2017 with a Steering Group to oversee the policy. An IPR was introduced at WHATWG mainly due to the fact that Microsoft could not join without one. As stated in 2007 by Chris Wilson, a member of Microsoft's Internet Explorer team:

> As an aside, I was asked personally to join the WHAT WG over a year ago. I had a back-and-forth discussion in email with several of the members of the WHAT-WG. I said something to the effect of 'I agree with your goals, all except standardizing error-correction*'. However, I also said 'you have no patent policy, and that makes it impossible for me to join.'

OASIS

The SDO OASIS does not stipulate an IPR policy but offers a number of IPR paths for standard adoption. This multi-track policy was introduced in 2005 which stipulated that WGs be charted under a set IPR rule. OASIS currently operates under four IPR tracks. Parties can either license a standard on a royalty-free basis, open source or on FRAND terms. With RF and FRAND, IPR is agreed in the final stages of the specification. Under the open source, licensing is agreed at the outset under Apache 2.0, BSD-3-Clause, CC-BY 2.0, Eclipse 1.0, and/or MIT licensing terms. Most standards in OASIS are RF with a decreasing number licensed under FRAND. This flexibility has attracted companies such as Apple, IBM, and Microsoft (Russell, 2011:171).

International Convergence in State Approach

Convergence is taking place between US and EU approaches towards SDO IPR policies, and SEPs in particular. 'Points of control' identified are: national acts,

[23] https://whatwg.org/ipr-policy.

court cases, competition policy, Business Review letters, workshops and opinions expressed by national governments and IGOs and informal pressure. Institutional actors are identified as larger states (the United States and the United Kingdom) and international organizations (the European Commission, European Court of Justice, and the World Trade Organization). Regional splits within the European Union can be identified due to historical national champion policies executed in Germany and France but transnational alliance has occurred between the United Kingdom, the European Union, and the United States (under the Obama Administration). It was pressure from the European Union and United States in particular which led to standards-development organization (SDO) change.

Following the governance literature, a key mechanism triggering change is systemic shock. The IoT can be seen as a systemic shock providing a window of opportunity for SDO reform. The IoT allows for the hooking up of devices across networks which were not historically connected to the Internet. Consumers are familiar with IoT devices in the home such as fridges and light switches that can be connected to mobile phones and for transport use such as connected cars. However, most IoT demand, according to a 2015 McKinsey Global Institute report, is from factories, smart cities, retail, and agriculture (McKinsey, 2015). Smart farming includes information on the consistency of soil, fertilizer application, weather, and monitoring animal health. Animals are monitored on activity, pulse, and body temperature. Farmers can predict potential illness earlier in this way. The McKinsey report estimated that the market for the IoT would be worth US$11.1 trillion per year by 2025 (McKinsey, 2015).

The identification of SEPs are often the only way to move agreement forward within SDOs. SDOs normally require their members to disclose SEPs and license them on either royalty-free or FRAND terms. However, the number of SEPs is sharply on the rise given the rapid development in the IoT where Cisco estimates that five 500 billion devices will be connected to the Internet by 2030.[24] As Contreras points out, 'complex technological products may implement dozens or even hundreds of standards—each of which may be covered by hundreds or thousands of SEPs' (2016:858). The IoT denotes a dramatic increase in demand for interoperability as sectors intertwine. The increase in SDO membership for operators seeking interoperability for their devices and products has led to an increase in disputes over patents which led this issue to be addressed at national and international levels (Figures 11.1 and 11.2). The

[24] https://www.cisco.com/c/dam/en/us/products/collateral/se/internet-of-things/at-a-glance-c45-731471.pdf.

Table 11.2 Most patents registering globally by company

Name	Number of patents		Country
	2016	**2018**	
IBM Corporation	8088	9,100	United States
Samsung	5518	5,850	South Korea
Canon KK	3665	3,056	Japan
Qualcomm Inc	2897	2,300	United States
Google	2835	2,070	United States
Intel	2784	2,735	United States
LG Electronics	2428	2,474	South Korea
Microsoft	2398	2,353	United States
Taiwan Semiconductor Manufacturing Company Ltd	2288	2,465	Taiwan
Sony Corp	2181	1,688	Japan
Apple	2102	2,160	United States

Source: IFI Claims Patent Services[a]
[a]https://www.ificlaims.com/rankings-top-50-2018.htm.

European Commission commissioned a study in 2014 which shows that SEPs result in a higher number of litigation cases than other patents (ECSIP, 2014).

The rise in patent disputes led state actors to become involved in SDO IPR policy setting. In 2010, this began with the European Commission issuing Guidelines on horizontal co-operation agreements[25] which mandate that SDOs have open membership, the procedure for adopting standards is transparent, there is no obligation for compliance, and there is access to FRAND terms. SDOs operating under this guidance are provided clearance of Article 101(1) so that standard agreements will not be viewed as anti-competitive and thus challenged under EU competition law. This was backed by the European Court of Justice which affirmed the Commission's right to intervene in SDO decision-making if these criteria are not met. Schellingerhout from the EU Directorate General for Trade stipulated that standard setting from a

[25] http://europa.eu/rapid/press-release_MEMO-10-676_en.htm.

competition law perspective implied that 'clarity on the intellectual property rights situation' be given and 'a commitment to license is given and respected' on behalf of SDO members (Schellingerhout, 2011:5). Under these conditions only the Commission could grant 'block exemptions' or provide 'safe harbour' for SDO decision-making.

In October 2012, an ITU Roundtable[26] on the effectiveness of FRAND took place where a number of proposals were discussed covering cross-licensing, injunctions, and third-party transfer. The US Department of Justice prepared six proposals for SDOs[27] which were largely adopted by the ITU. This was followed by a number of decisions by the US Supreme Court which have curtailed patent use[28] and cracked down on patent trolls by limiting conditions under which patent cases can be brought.[29] This is a different track to past support of patent policies within the United States.[30]

This was followed in 2014, by the European Commission's Communication on Patents and Standards: a modern framework for IPR-based standardisation which focused on the lack of transparency regarding SEPs, improvements to the patent declaration system, promotion of patent pools, and transfer of SEP ownership. Then, in 2015, the European Court of Justice (ECJ) ruled on the *Huawei v ZTE* case which dealt directly with SDO policies (Grasso, 2018). In the ECJ case, Huawei held an SEP for the Long Term Evolution (LTE) standard developed within ETSI. This began in 2013 when Huawei notified ETSI that it held the patent and was committed to licensing on FRAND terms. The SEP was utilized by the company ZTE but would not agree to the licensing arrangements proposed by Huawei, which then sought an injunction in a German court (Grasso, 2018:96). ZTE counter-challenged claiming that Huawei was exercising abuse of dominant position. The ECJ used the case to outline lengthy restrictions on SEPs and FRAND. These developments provoked review and change of IPR policy in each SDO with active steer from state organizations.

The adoption of the IEEE's 2016 policy was particularly contentious (Kanevskaia and Zingales, 2016). In 2015, the US Department of Justice produced a Business Review[31] at the request of the IEEE. Business review letters are

[26] http://www.itu.int/en/ITU-T/Workshops-and-Seminars/patent/Pages/default.aspx.

[27] https://www.justice.gov/atr/file/518951.

[28] 2016 *Samsung v Apple*; 2016 *Halo Electronics v Pulse Electronics*; 2015 *Commil v Cisco*; 2014 *Alice Corp. v CLS Bank*; 2014 *Ericsson v D-Link*.

[29] See: https://www.nytimes.com/2017/05/22/business/supreme-court-patent-lawsuit.html, and explanation by the EFF at: https://www.eff.org/deeplinks/2017/05/supreme-court-ends-texas-grip-patent-cases.

[30] For example, the FTC lost its case against Rambus in 2008 in the US Court of Appeals.

[31] https://www.justice.gov/file/338591/download.

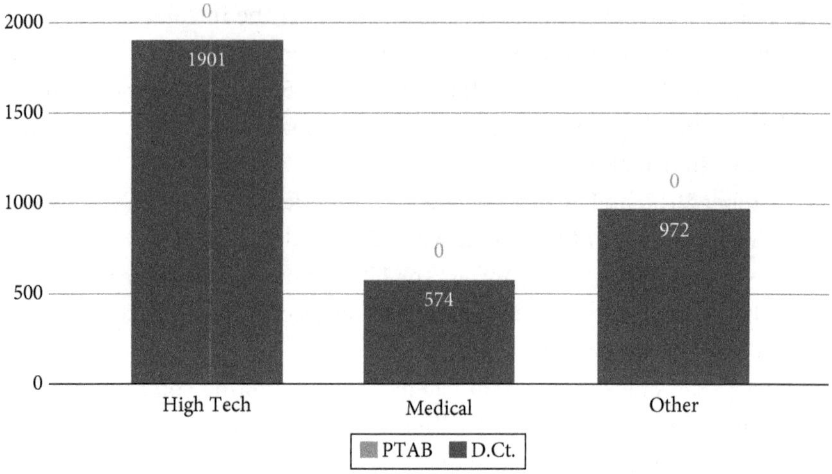

Figure 11.3 Patent litigation by industry in 2018 by sector at Patent Trial and Appeal Board (PTAB) and the District Court levels.
Source: IPlytics GmbH[a]

[a] https://www.iplytics.com/report/sep-auto-industry-5g/. IPlytics—Tim Pohlman. More up-to-date statistics are available from https://www.iplytics.com, with information on standard essential patent owners which include more granted and non-granted SEPs.

issued to companies by the Department of Justice pursuant to the Deparments's Business Review Procedure to provide guidance about the legality of proposed conduct under antitrust laws. A number of companies within the IEEE, which were dependent on IEEE patents to build products, used this document to push for a change including Apple, Cisco Systems, Intel, Microsoft, Samsung, and Verizon. The resulting 2016 IEEE policy[32] states that patent holders must license their patents for 'any compliant implementation' which means any product under discussion at the IEEE. A Letter of Assurance (LoA) must be submitted stipulating the submitter's position on ownership, enforcement, and licensing of SEPs, a ban against injunctions opposing SEPs, and a 'reasonable rate'. The reasonable rate is the most contested issue as it requires that:

- The value that the functionality of the claimed invention or inventive feature within the Essential Patent Claim contributes to the value of the relevant functionality of the smallest saleable Compliant Implementation that practices the Essential Patent Claim.
- The value that the Essential Patent Claim contributes to the smallest saleable Compliant Implementation that practices that claim, in light of

[32] http://standards.ieee.org/develop/policies/bylaws/sect6-7.html#loa.

the value contributed by all Essential Patent Claims for the same IEEE Standard practised in that Compliant Implementation.

- Existing licenses covering use of the Essential Patent Claim, where such licenses were not obtained under the explicit or implicit threat of a Prohibitive Order, and where the circumstances and resulting licenses are otherwise sufficiently comparable to the circumstances of the contemplated license.[33]

The IEEE managing director Konstantinos Karachalios stated that the policy would introduce 'more clarity and predictability for patent holders, implementers and users around technologies that will be included in our standards going forward, such as the IEEE 802.11 family of Wi-Fi standards'.[34] In his analysis of the IEEE, Katznelson noted a rapid decline in non-duplicate licensing LoAs for IEEE standards (2016:9). After adoption, however, Qualcomm, Ericsson, and Nokia refused to license their SEPs under the new rules. Alcatel-Lucent, Ericsson, and Qualcomm appealed to the IEEE to remove accreditation from ANSI (Kanevskaia and Zingales, 2016) but ANSI turned the request down. Subsequently, a number of companies operating within the IEEE formed the Innovation Alliance which, according to its website, 'believe in the critical importance of maintaining a strong patent system that supports innovative enterprises of all sizes'.[35]

In January 2017, ANSI issued a new document on Essential Requirements: Due process requirements for American National Standards.[36] In the document, ANSI recognizes the Technical Barriers to Trade (TBT) Code of Good Practice which adopts WTO norms. This was followed by a 2017 US Federal Trade Commission (FTC) report on patent assertion entities.[37] Shortly afterwards, the FTC accused Qualcomm of 'unlawful maintenance of a monopoly in baseband processors, semiconductor devices that enable cellular communications in cell phones and other products' and took the company to court.[38] The claim was that Qualcomm utilized its market position to exact favourable licensing terms from mobile phone companies, and, in particular, 'Qualcomm

[33] https://standards.ieee.org/about/policies/bylaws/sect6-7.html.

[34] https://www.computing.co.uk/ctg/opinion/2425663/current-patent-policies-are-houses-without-foundations-in-regions-of-earthquake-risk.

[35] https://innovationalliance.net/.

[36] https://share.ansi.org/shared%20documents/Standards%20Activities/American%20National%20Standards/Procedures,%20Guides,%20and%20Forms/2017_ANSI_Essential_Requirements.pdf.

[37] https://www.ftc.gov/news-events/press-releases/2016/10/ftc-report-sheds-new-light-how-patent-assertion-entities-operate.

[38] https://www.ftc.gov/news-events/press-releases/2019/05/statement-federal-trade-commission-bureau-competition-director.

has engaged in exclusionary conduct that taxes its competitors' baseband processor sales, reduces competitors' ability and incentive to innovate, and raises prices paid by consumers for cell phones and tablets'. In May 2019, the FTC won the first part of the case in a Californian federal court. The judge ordered Qualcomm to change its licensing terms.

Action in the United States was flanked by activity in Europe. The European Commission published three reports on IPR in 2015.[39] Then, in its 2016 Communication on ICT Standardisation Priorities for the Digital Single Market,[40] the European Commission referred to standards for the first time 'as a strategic instrument to EU industrial policy'. Its subsequent 2017 Communication Setting out the EU approach to Standard Essential Patents[41] focused on the IoT. Citing the 2015 McKinsey report, the European Commission argued that 40% of the capacity for IoT systems could not be achieved without interoperability. For this reason, the European Commission asserted that, 'there is therefore a need for a clear, balanced and reasonable policy for Standard Essential Patents in the EU with the aim of contributing to the development of the Internet of Things and harnessing Europe's lead role in in this context'.

The Communication continued with a set of key principles to ensure predictability for SEPs within SDOs. Under Principle 1, the Commission stipulated that transparency for SEPs should be increased via accessibility in SDO databases, tools for licensing negations, and the updating of SDO declarations. Principle 2 outlines general principles for FRAND licensing terms for SEPs. Here the Commission mandated the publication of clear principles, non-discrimination, the establishment of transparency patent pools within the scope of competition law, and increasing FRAND expertise with a European Commission expert group to monitor 'sound IP valuation and FRAND'. It insisted that FRAND be established on the current value of patented technologies and not future estimates. Under Principle 3, the Commission requested that a predictable environment for SEPs be created. This incorporated the decision in the *Huawei v ZTE* case where the ECJ stated that injunctions cannot be

[39] Comino, Manenti, and Thumm (2015) 'Intellectual Property and Innovation in Information and Communication Technology', Joint Research Centre. Available at: http://publications.jrc.ec.europa.eu/repository/bitstream/JRC97541/jrc97541.pdf; Pentheroudakis and Thumm (2015) 'Digital Single Market: The Role of Patents', Joint Research Centre. Available at: http://publications.jrc.ec.europa.eu/repository/bitstream/JRC96728/jrc96728.pdf; and Ménière (2015) 'Fair, Reasonable and Non-Discriminatory (FRAND) Licensing Terms', Joint Research Centre. Available at: http://is.jrc.ec.europa.eu/pages/ISG/EURIPIDIS/documents/05.FRANDreport.pdf.

[40] https://ec.europa.eu/digital-single-market/en/news/communication-ict-standardisation-priorities-digital-single-market.

[41] https://eur-lex.europa.eu/legal-content/EN/ALL/?uri=CELEX%3A52017DC0712.

sought against users willing to enter into licenses on FRAND terms. If a case is brought, rather than seeking injunctions in court, the Commission encourages alternative dispute resolution via the Unified Patent Court which is soon to be established for European patent disputes.[42] The Commission Communication stipulated that patent assertion entities, those which do not hold patents as such but enforce on behalf of SEPs holders, should be subject to the same rules as SEPs holders. Finally, under Principle 4, the Commission encouraged open source software and open source standardization.

Two further reports were issued by the European Commission in 2017. The first provided an overview of patent assertion practices in the context of the IoT.[43] The report showed that there is higher litigation over patents in the United States than in the European Union with high representation from the telecommunications sector and most litigation within the European Union taking place in Germany. The second offers an overview of how FRAND licensing provides forty case studies in the European Union, the United States, and Asia with economic methods to assess FRAND (Pentheroudakis and Baron, 2015). These were followed with two research projects. The first dealt with the governance of standards developing organizations and IPR in IoT with seventeen case studies: the Association Française de Normalisation (AFNOR, ANSI, the European Committee for Standardization and European Committee for Electrotechnical Standardization (CEN-CENELEC), the Deutsches Institut für Normung DIN, Digital Video Broadcasting (DVB), the European Computer Manufacturer's Association (ECMA), TSI, the International Electrotechnical Commission (IEC), IEEE-SA, IETF, the International Organization for Standardization (ISO), ITU, Joint Electron Device Engineering Council (JEDEC), Standardization Administration of the People's Republic of China (SAC), Telecommunications Standards Development Society, India (TSDSI), VITA, and the W3C).[44] The second focused specifically on the interaction between open source software and FRAND in standard setting.[45]

In the context of IoT, the Commission highlighted a number of areas that needed to be addressed: an imbalance in information about SEPs and the applicable regulations particularly for new market players unfamiliar with SEPs

[42] The Unified Patent court (https://www.unified-patent-court.org/) is to be established under the Unified Patent Court Agreement (16351/12) and EU Regulations 1257/2012 and 1260/2012.

[43] Patent Assertion Entities in Europe: Their impact on innovation and knowledge transfer in ICT markets https://ec.europa.eu/jrc/en/publication/eur-scientific-and-technical-research-reports/patent-assertion-entities-europe-their-impact-innovation-and-knowledge-transfer-ict-markets.

[44] https://ec.europa.eu/jrc/en/publication/eur-scientific-and-technical-research-reports/making-rules-governance-standard-development-organizations-and-their-policies-intellectual.

[45] https://ec.europa.eu/jrc/en/science-update/jrc-study-interaction-between-open-source-software-and-frand-licensing-standardisation.

in SDOs such as car manufacturers; predefined licenses per module (e.g. chipsets) versus licensing requests for standards already implemented (often by Patent Assertion Entities (PAEs)); percentage fees versus (fixed) per-unit fees; the double dipping of component maker and end product producer; that SEO owners prefer licensing to the end product maker versus end product makers (e.g. car manufacturers) requiring that component makers deliver components without IPR restrictions (and that concessions on both sides were needed); 'user-based licensing' for example based on value of car, chip use in smart meters (which provided little information) versus in cars/people (which provided lots of information); patent exhaustion and the refusal to license if the licensee exhausts the inventor's right; and control of end use of technology via 'field of use restrictions'.

Conclusion

Clearly, SDOs have found the definition of IPR policies problematic. In the early 1980s, SEPs were not an issue as the ICT sector was mainly dominated by national incumbents, particularly in Europe with Alcatel, Ericsson, and Siemens but also with large corporations in the United States such as GEC. Only with deregulation, followed by globalization, did IPR within SDOs become important first with standards for mobile phones such as CDMA (Qualcomm), theGlobal System for Mobile Communications (GSM)[46] and WiFi (the 802.11 family of standards). The rise in patent disputes climbed in the 2000s and is expected to rise again in the 2010s due to the advent of the IoT. This phenomenon has led to windows of opportunity for SDO reform in intellectual property rights. Actions by states, particularly the United States and within the European Union, has led actors within SDOs to push for more open and flexile IPR approaches. In this respect, state steer, as a macro institutional mechanism, has influenced decision-making within the politics stream.

This has been flanked by greater use of open standards. As Halpin notes, this is particularly important for inclusion of civil society and consumer groups into standard setting given that 'academics and activists can participate, although in practice participation is limited due to time, travel, and lack of academic incentives' (2016:1). Russell notes that open standards also present a bureaucratic incentive as the licensing process is 'costly and time-consuming'

[46] Originally developed by Motorola, GSM was adopted by ETSI in 1990. Subsequently Ericsson, Nokia Alcatel, and Siemens developed patents for the standard.

(2011:162). This is notable progress within the W3C, IETF, and the IEEE on IPR as they are members of Open Stand which advocates open standards.[47] However, although the IEEE has made some progress with reform and agreement over key patent-free standards such as USB and (initially) Bluetooth, there is an abundance of SEPs licensors within the mobile communications market particularly within ETSI, the Third Generation Partnership Project (3GPP), and the ITU-T.

The IoT has introduced new players into SDO fora which were seeking interoperability for devices. Indeed, there are a number of new players on the market due to the IoT but also new standards bodies, for example the 2014 established Open Automotive Alliance (OAA) for standardization between automotive manufacturers and technology companies for the use of Android in automobiles and the Agricultural Industry Electronics Foundation (AEF), the Alliance for Internet of Things Innovation, the Open Connectivity Foundation (OCF), Calypso (open standards for ticketing), and the CAR 2 CAR Communication Consortium. There is also a push from the European Commission particularly towards the use of open source standards. These have been taken on board in the IoT sector again with the establishment of new fora such as the Open Connectivity Foundation, Arduino, and the Avnu Alliance.

[47] https://open-stand.org/.

12

Conclusion: SDO Decision-making and the Public Interest

Standard setting, which determines interoperability, is at the heart of the functioning of the Internet. The findings from this book suggest that standards-development organization (SDO) coordination is the most efficient method, affording a higher level of expertise and technical knowledge than standards determined by state agencies. Implementation of SDO standards is voluntary and implementation costs are removed from the shoulders of the public sector. The Internet governance literature has concentrated on representation of the public interest in international organizations. However, most decision-making about the Internet occurs within self-regulatory fora such as the Internet Engineering Task Force (IETF), the World Wide Web Consortium (W3C), and the Institute of Electrical and Electronics Engineers (IEEE). Members of these organizations are volunteers, normally seconded from the corporate sector, with no formal state representation (Alvestrand and Lie, 2009). As Littoz-Monnet points out, and as many of the chapters of this book have shown, experts 'do not necessarily act as cohesive communities' (2017:16). This is particularly the case if they are mainly representative of large global corporations in high-stakes industrial sectors like those invariably associated with the Internet. Global standards-making is also subject to significant concentration. Although there are thousands of standards bodies, 'the 20 largest SDOs (produce) 90% of the standards' (OECD, 2011:254).

Extant literature queries whether business regulation is still possible given the rise of self-regulation and the political power of global corporations, law firms, and active-knowledge politics which dominate transnational legal orders (Prosser, 1997, 2010a, 2010b, 2011; Delazay and Garth, 1998). The global governance literature claims that SDOs undermine national democratic processes and it documents the erosion of state power and public sector actor influence due to globalization and the move towards self-regulated systems of governance (Braithwaite and Drahos, 2000; Picciotto, 2011; Lundqvist, 2014). This 'rise of the unelected' (Vibert, 2007) raises the important question of the

Global Standard-Setting in Internet Governance. Alison Harcourt, George Christou and Seamus Simpson, Oxford University Press (2020).
DOI: 10.1093/law-oso/9780198841524.001.0001

representation of the public interest at the international level. Although the literature is expansive on the interaction of corporate sector actors within international self-regulatory fora (Braithwaite and Drahos, 2000; Hall and Biersteker, 2002; Drezner, 2004; O'Neill, Balsiger, and VanDeever, 2004; Farrell, 2006; Jupille, Mattli, and Snidal, 2013), it has not yet addressed the actions of civil society and those within SDO fora who promote public interest goals. This book has addressed this gap.

The book's findings largely confirm conclusions in the global governance literature which inform us that corporate actors dominate international self-regulatory fora with significant steer from key member states, particularly the United States but also the European Union. It demonstrates that, even in areas where the state showed long-term historical dominance, policy matters are spilling over into the purview of standards fora. Drezner illustrated how states still dominated in the realm of content regulation, intellectual property rights (IPR), and privacy rights as these were areas of high political conflict (2004:483). Princen showed that high salience triumphs over low salience issues. States have also driven the work more than civil society (2007). However, Chapters 9 and 10 of the book find that spillover to the international level has removed the role of the state and reinstated strong private governance in these areas. Spillover has increased the role of the private sector in these aspects of Internet governance and largely removed it from public scrutiny.

However, the book also showed a more nuanced pattern emerging with increasing inclusion of public interest goals within SDO fora. In this respect, transnationalism is evidently opening new spaces for societal actors within global governance (Cerny, 1995; Risse-Kappen, 1995; Strange, 1996; Scholte, 2000). This chapter concludes by exploring areas and opportunities for public interest representation. In turn, it addresses the utility of the book for scholars working on global governance, self-regulatory for, and Internet governance.

The first half of the chapter will analyse findings of the empirical cases presented in the book in order to explain which mechanisms trigger standards development as a form of policy adoption envisioned through the multiple streams (MS) framework. In the next section, the case studies are analysed on a comparative basis, explaining and identifying the processes under which SDO rules of interaction are established at the international level, whose interests are represented in SDO fora, and how. They highlight the occurrence of policy entrepreneurship, forum shopping, and coupling uncovered in the preceding chapters. Each case described policy problems access the agenda, how and where constellations of actors carry out their search for solutions, and how the selection of options is biased and contingent on a range of internal and external factors. The final part of the chapter explores additional frameworks for SDO regulation where the areas for public interest consideration might occur. These are opportunities for inserting public interest considerations into international and national Acts, certification programmes, and the move towards open source solutions for Internet management.

Comparing the Cases: What Triggers Policy Adoption?

The book advanced work on the multiple streams (MS) framework by applying it to decision-making in non-state environments, namely SDOs, which have long been dominated by private actors. The decision-making lens of the framework is most suitable to the book's investigation as it allowed us to identify mechanisms, and constituent factors within them, which lead to policy decisions. Self-regulation is a form of governance that relies on decision-making processes in the absence of centralized control. It assumes the absence of clear objectives, other than quite often to keep government out, fluid participation, and sometimes a lack of clear internal rules on the operation of self-regulatory fora. The MS perspective fits with these conditions as it makes precisely these assumptions.

The first part of this section analyses how macro mechanisms have been able to steer agendas by constraining problem recognition and exploiting temporal events which determine the direction of SDO policy formation. It explains how MS policy streams, windows, and entrepreneurs in this context provide conditions under which internal and external actors are able to influence the direction of decision-making. The main external actors identified are states and digital rights and consumer groups which often have no formal involvement in SDO decision-making. Strategic mechanisms identified in the main empirical case study chapters include mobilization by these actors through conflict (e.g. public campaigns), competition court cases, and the manipulation of both subsidiarity and multi-level governance.

The second part of the section turns to micro mechanisms for policy change. Actors choose arenas based upon cognitive, strategic, and legitimacy mechanisms. Cognitive mechanisms relate to the framing, problem redefinition, and policy narratives of an issue. This level is more efficient than the strategic mechanism because it determines who coordinates a given action. The chapter explains how actors use normative arguments to claim that one arena is more appropriate than another, whether this be national, international, legislative, or self-regulatory.

These legitimacy mechanisms shift arenas and stack the deck in that they filter types of influence and the resources available to actors in their choice of policy forum and standard to adopt. Legitimacy mechanisms also privilege one type of values, norms, and behaviours over others, providing causal influence on policy outcomes. Choosing an arena can also create inertia. Once a group enters the policy game in one arena, it is costly to shift to a different area. However, it should be noted that forum shopping does not just take place between SDOs but also 'within' SDOs and different working groups. Overall, both sections in this chapter offer an explanation of the framing and coalition-building strategies of important actors, their ability to influence the agenda, and the mechanisms that lead to eventual decisions.

Macro Mechanisms for Policy Change

As outlined in Chapter 2, the book identified different mechanisms which empowered actors within SDO fora: macro (exogenous and institutional) and micro (collective and individual). Each chapter then observed how these mechanisms work within the different MS streams. The first macro mechanism is exogenous. Chapters 4, 8, 9, and 11 identified how systemic shocks and technological change can act as focusing events which strengthened the ability of certain internal actors to alter SDO agendas.

Cairney explained that policy entrepreneurs are successful when 'exploiting crises or events, selecting the measures that produce the most supportive evidence of a problem' (2018:201). The Snowden revelations reshaped the debate over protocol development within the IETF. Chapter 4 showed how the 2013 Snowden revelations created more urgency in the problem stream of solving security flaws in TLS. These revelations created a seismic shock and a great deal of soul-searching for the IETF when it was revealed that the US National Security Agency (NSA) and the UK Government Communications Headquarters (GCHQ) were capturing vast amounts of Transport Layer Security (TLS) traffic. From 2014, different solutions were discussed for the IETF's TLS 1.3 standard, which sought to replace the more problematic TLS 1.2.

Quick User Datagram Protocol (UDP) Internet Connection (QUIC) obscures traffic over mobile communications networks making traffic management and data collection on unencrypted traffic impossible. It also obfuscates pervasive monitoring efforts, notably by national governments. This challenges requests by governments for operators to pass monitored traffic on to intelligent agencies for surveillance purposes. If mobile network operators cannot 'see' the data, they cannot comply. For this reason, QUIC was supported strongly by digital rights groups externally and those few active internally to the IETF, namely the Electronic Frontier Foundation (EFF), Center for Democracy & Technology, Article 19, and American Civil Liberties Union (ACLU) representatives. It was also arguably appreciated by technology companies tired of the increasing number of government requests. In this respect, the Snowden revelations presented a window of opportunity to couple an existing solution (Google's work on QUIC) within the problem stream (TLS security flaws).

QUIC was originally an experiment designed by Jim Roskind at Google. The protocol was introduced mainly to reduce latency. As explained in Chapter 4, it replaces TLS over Transmission Control Protocol (TCP) with QUIC over User Datagram Protocol (UDP). This reduces the 'handshaking' time between servers. It also improves loss recovery and congestion and removes breaks in transmission. Importantly, everything is encrypted so there can be no intervention by third parties when packets are transferred. This is ensured via a cryptographic handshake. Google chose to standardize QUIC not because it was seeking a solution to TLS security but because Google is a leading global provider of data servers. When transferring data between Google-owned servers, using QUIC is simple, but moving packets to other companies' servers would mean that these companies would also have to be using QUIC. Therefore, Google was seeking standardization and interoperability. It was incentivised to improve its market position by providing servers with greater speed.

Competing positions arose amongst actors, namely companies supporting QUIC and those against it. Normally this opposition would have been insurmountable and it explains to a great extent status in decision-making over replacement of TLS 1.2 up until Snowden. Chapter 4 detailed how Google's QUIC standard created an opportunity to address the issue of enhanced security (encryption) through ongoing work on TLS. The chapter explored the different solutions proposed within the policy stream and how these have played out in relation to building and reaching a consensus within the IETF on TLS 1.3 and QUIC. The overall consequences of Snowden on IETF were apparent in QUIC discussions along with the parallel development of the Pervasive Monitoring Request for Comments (RFC) 7258. Importantly, the adoption of QUIC was supported by digital rights groups and a hybrid solution to privacy and data protection was promoted, even though contested.

The discussion over QUIC did not change the overall ideational environment in which engineers work within the IETF. The IETF itself did not want to adopt QUIC as a replacement for TLS but develop TLS 1.3 as a cryptographic infrastructure for protocols using UDP such as QUIC. Enabling QUIC was an important decision, however, given that the Google footprint is wide, with Netflix, Akamai, Snapchat, and iOS, having adopted QUIC by 2017. By mid-2018, QUIC was in use by 46% of the domain name system (DNS) names space (Rüth, Poese, Dietzel, and Hohlfeld, 2018). However, Snowden's impact empowered wider actors within the IETF, particularly digital rights and consumer groups, which enabled coupling of a new solution for QUIC and eventually TLS 1.3 based upon a more guarded approach to securing internet traffic.

Chapter 11 explained how the Internet of Things (IoT) acted as a systemic shock, opening a window of opportunity for SDO reform in intellectual property rights (IPR). This opportunity strengthened external state actors who pushed SDO decision-making towards a more open and flexible IPR approach. State steer was identified as a macro-institutional mechanism in the politics stream. The IoT introduced new players into SDO fora which were seeking interoperability for devices. The most familiar uses of IoT are products in the home such as alarms, in-home cameras, fridges, lights, speakers, and heating thermostats which can be controlled from mobile phones, computers, or home hubs such as Google's Nest Hub and Amazon's Alexa. However, most IoT growth has come from agriculture, factories, smart cities, and retail sectors. The increase in the number of standard essential patents (SEPs), resulting from the convergence of different sectors' need for interoperability, threatened to increase the cost to the consumer enormously.

The chapter described different solutions within the problem stream, which ranged from actors supporting a more proprietary approach and those advocating a more open approach to IPR. External exogenous pressure empowered internal actors advocating more open approaches to IPR. SDOs changed IPR policies to include greater transparency, commitments to reasonable and non-discriminatory licensing, and to increasing embrace of open source solutions. This was notably derived through soft governance. Although competition decisions and court decisions had their role to play, the main 'points of control' took the form of US Department of Justice Business Review letters, workshops, and opinions expressed by national governments and International Governmental Organizations (IGOs) as well as other forms of informal pressure. The main institutional actors were larger states (the United States, the United Kingdom, and Germany) and international organizations (the European Commission, European Court of Justice, and the World Trade Organization (WTO)).

The European Commission acted as a policy entrepreneur using IoT to identify SEP licensing issues. It also drew attention to the fact that SDOs have preferences for licensing SEPs to end product makers, such as car manufacturers, representing an imbalance of actors within the fora. The solutions proposed were: greater transparency in information about SEPs; predefining licenses per module; introducing fixed fees within SDOs; preventing 'double dipping' of component makers and end product producers; and to make licensing 'user based', based upon the value of the final product (such as a car or chip used in smart metres), thereby consumer oriented.

External actors have largely been responsible for driving SDO policies governing state surveillance. Chapter 9 showed how the market-driven search for standard interoperability has been hampered by state security interests. It identified the problem as state surveillance policies which can affect privacy and data protection for citizens. The framing of the problem, however, had more to do with intervention in the market operation of companies. The problems identified by SDOs included state restrictions on cryptography research; the interception of TLS traffic by security agencies; the development of protocols by intelligence agencies; and concerns over 'back-door' technologies. The Snowden revelations triggered a window of opportunity for the IETF to classify pervasive monitoring as an attack. The post Snowden work on how to mitigate against pervasive monitoring in the design of protocols did not trigger an overall ideational change within the IETF. However, it empowered existing actors within these fora pressing for solutions to mitigating state surveillance in protocol development.

This main work took place within the IETF in the securing of servers and Internet Protocol (IP) identification, Session Initiation Protocol (SIP), Post Office Protocol 3 (POP3), Internet Message Access Protocol (IMAP), Simple Mail Transfer Protocol (SMTP), and Media Access Control (MAC) protocol and Hypertext Transfer Protocol (HTTP) cookies. The Internet Research Task Force (IRTF) established the Human Rights Protocol Considerations Research Group (HRPC) to consider integration of key principles into standard development as defined in the Universal Declaration of Human Rights (UDHR) and the International Covenant on Civil and Political Rights (ICCPR). As discussed in the chapter, although this goes some way to the recognition of fundamental human rights in protocol development, it was not well received within the IETF. It did, however, create a space for the adoption and roll out of the HTTP status code 451 which signals the take down of content on the web via legal measures. Ultimately, however, SDOs are governed by the overarching legal architectures within which they operate which constrain decision-making.

Chapter 9's conclusion points to the strengthening of formal legal measures which enable state surveillance and monitoring.

Two contrasting factors led to a window of opportunity for a new solution to historical contestation over spectrum space: the fact that many users in rural and developing economies remain unconnected to the Internet and, by contrast, the promise of socio-economic transformation through extensive machine-to-machine (M2M) communication as part of the burgeoning IoT. Chapter 8 explains how technical standards making was able to present dynamic spectrum access as a policy solution to the problem of intensifying use of television white spaces (TVWS). Yet, political contestation between a range of interests from broadcasting (digital terrestrial television (DTT) providers, programming making and special events (PMSE) interests, audience groups) and telecommunications (licensed operators, new TVWS entrants) led to under-developed new TVWS markets despite relatively strong state steer in favour of introducing dynamic spectrum access- (DSA) enabled unlicensed access to TVWS for new service providers.

Both US and UK regulatory bodies set in train what turned out to be contested policy processes to broaden TVWS use. Technical standards-making in the IEEE, in particular, provides the basis for equipment vendors to develop database and wireless communication equipment necessary to operate tiered priority access and use of the spectrum. Yet, evidently strongly contested positions in consultation exercises launched by regulators in the United Kingdom and the United States, as well as slower than expected deployment and take-up of new services, meant that TVWS is yet to fulfil its potential. Technical standards underpinned DSA as it provided a techno-functional—though not a commercial or social policy solution—to the TVWS policy problem.

Micro Mechanisms for Policy Change

As outlined in Chapter 2, micro mechanisms are characterized first via collective change, namely the creation of constituencies and boundaries, gradual change, layering, and policy drift, and second via individual change through actor behaviour and learning, changing perceptions, and the social construction of roles (Falleti and Lynch, 2009). Coupling comes about when the context for change remains stable but the social construction of roles governing actor behaviour changes internally. Chapters 5, 6, 7, and 10 examined case studies for evidence of actor-led ideational change based upon various articulations of micro mechanisms (see Table 12.1). The conclusion in each case is that

Table 12.1 Case study summary

	problem stream	policy stream (solutions)	politics stream (key actors and entrepreneurs)	mechanism (enabling coupling)
Chapter 4	security flaws in TLS	• QUIC • TLS 1.3	• corporations • digital rights groups	macro exogenous
Chapter 11	proprietary standards	• more open IPR policies • proprietary standards	• states • corporations • open source groups	macro institutional
Chapter 9	state surveillance	• pervasive monitoring policy • transparency measures • technical solutions to surveillance	• engineering community • digital rights groups • security agencies	macro exogenous
Chapter 8	efficient white space spectrum	• dynamic spectrum access	• states • corporate groups	macro institutional
Chapter 5	HTML development	• HTML 5.2 • XML • EME	• browser companies • rights holders • digital rights groups	micro collective
Chapter 6	efficient local area network wireless use	• 802.11ax	• Special Interest Group • states • consumer	micro collective
Chapter 7	consumer privacy	• technical solutions • DNT	• corporate groups • consumer groups	micro individual
Chapter 10	consumer privacy	• technical solutions	• corporate groups • digital rights groups	micro collective

although ideational clashes can be observed, they did not lead to overall actor-led ideational change within SDO fora. Indeed, decision-making which is not affected by external focusing events continues to be driven by private sector concerns based on market drivers, albeit not always harmoniously agreed.

Streeck and Thelen refer to categories of layering, drift, conversion, and exhaustion where change may take place (2005:31). They point to mechanisms of defection, differential growth, deliberate neglect, redirection/reinterpretation, and depletion which may lead to institutional change. Chapter 5 explains empirically how layering occurred in the development of the HTML standard. Stasis in HTML development within the W3C and layering with its subsequent development within the Web Hypertext Application Technology Working

Group (WHATWG) acted as micro mechanisms triggering windows of opportunity for coupling different solutions to the problem of web architecture development.

A problem stream was created when the W3C began to develop XHTML to replace HTML. When XHTML 2.0 was introduced by the W3C in 2002,[1] it was incompatible with the older HTML/WHTML versions. This resulted in the creation of an alternative SDO forum, WHATWG, which furthered work on the older HTML. This led to layering where a new standard was introduced alongside an existing one. A coupling of the policy stream (WHATWG's solution) occurred when WHATWG's HTML 2.0 was adopted by IBM, Microsoft, and Apple. WHATWG acted as a policy entrepreneur in convincing the engineering community to adopt the standard. This was because there was a strong match between the problem of non-compatibility and at the same time strong support from the commercial sector for the solution. In this case, an external focusing event did not occur. The technical feasibility of the solution was enough to couple the streams. The desire to create a better web, both for purposes of efficiency and consumer protection, has been never been realized mainly due to the need for web developers across the Internet to adopt the new system all at once. The development of Internet browser standards, with all their recognized flaws, continues to be developed in WHATWG based on the preferences of the main browser companies.

The second part of the chapter showed how the loss of HTML and Document Object Model (DOM) development to WHATWG resulted in a shift in focus of the W3C to development of the semantic web, the web of devices, and web and television standards. This opened a window of opportunity for a move of technical solutions for digital rights management (DRM) to the W3C. These technologies, such as watermarking and fingerprinting, rights expression languages, and application programming interfaces (APIs) for web browsers, allow clips and mash-ups of online music, videos, and text to be easily traced and quickly blocked or automatically taken down when copyright infringement occurs.

EME's adoption by the W3C in 2017 in particular was highly controversial because it facilitates the mandatory use of proprietary digital rights management technologies in browsers. The adoption of Encrypted Media Extensions (EME) demonstrated that the shift away from agreements in international treaty-based to technical standards fora renders key developments in the Internet's architecture largely invisible to consumer groups. The conclusion

[1] http://www.w3.org/TR/2002/WD-xhtml2-20020805/.

here is that when decision-making is removed from the level of the state or bilateral treaty agreements and reframed in technical terms, representation from consumer groups and the wider public is low.

Chapter 6 focuses on the increasingly popular—but thus crowded—environment of the 2.4 and 5 GHz unlicensed wireless local area network (LAN) spectrum bands and the core problem of creating high-quality communications services amidst ever-increasing use of this limited space. The IEEE 802 family of standards spawning WiFi services underpinned the service environment in question, although being open entry in character, the unlicensed local wireless environment bands accommodated increasingly licensed mobile communications operators alongside unlicensed WiFi Internet access providers. The chapter shows how the 802.11ax standard was developed as a solution to the problem of increasing service quality in a very congested communications space. This policy endeavour was buffeted by the highly disruptive DensiFi Special Interest Group, involvement of which acted as an internal mechanism which delayed, though did not scupper, the development of the standard. The case of 802.11ax also showed evidence of a highly contested political stream where rival standards making interests from the licensed spectrum world presented the long-term evolution (LTE) standard as a challenger to 802.11ax in terms of providing the necessary quality of service in the unlicensed bands. The political stream was intensified by evidence of strong public consumer preferences for mobile devices able to deliver content-rich, speedy services through seamless switching between licensed cellular and unlicensed WiFi environments. A third external factor in the political stream was legislative change in the European Union and the need for 802.11ax to be ready in order to be covered by a modified ETSI standard developed in response to the EU's Radio Equipment Directive. The development of 802.11ax is illustrative of the high-stakes nature of technical standards-making for the Internet and how internal mechanisms and external pressures exercised in the political stream of the standards-making process can influence the course and outcome of a standard's development in response to an identified problem stream.

Chapter 7 demonstrates how a problem stream emerged when it became evident that self-regulation of consumer privacy was failing. A window of opportunity arose when the US Federal Trade Commission (FTC) endorsed the development of the Do Not Track standard (DNT). This coupled the problem stream and the politics stream by focusing W3C efforts on finding solutions to unsanctioned online tracking. Different solutions were pushed by advertisers and publishers, Google (via a technical solution), and privacy advocates. Ultimately the solution preferred by corporate actors succeeded mainly due to stagnation and policy drift within the W3C resulting in the agreement over DNT but this was largely rendered redundant due to the lack of consumer take-up. The conclusion is that, even though there is evidence of ideational conflict, there was no ideational change within the W3C as a whole in favour of the public interest. Without external focusing events and, in particular, state action, the issue of consumer privacy is unlikely to be addressed within SDO fora.

Chapter 10 details the spillover of policy domains formally defined at the level of the state to SDOs with case studies on consumer privacy and copyright. The first part of the chapter explained how the problem of consumer privacy was identified early on within SDO fora. It is recognized that devices are fingerprinted and collect and sell users' data without their knowledge. This has led to increased work to mitigate browser fingerprinting leading to an IETF RFC 7230 in 2014 and W3C Candidate Recommendation in April 2016. The second half of the chapter covers accounts of digital rights activism on company behaviour.

These efforts, however, occurred around the fringes of main SDO work and are mainly initiated and carried out by civil society, academics, and specialist groupings. Indeed, it is clear from interviews that the problems of consumer privacy have long been recognized by SDO participants and are based on underlying societal norms concerning the right to transparency and privacy. Public campaigns by the EFF, a *New York Times* investigation, and recent action by the FTC and the UK parliament are resulting in more public recognition of the problem. However, serious attempts to address the problem by SDO fora cannot come about without direct state action in favour of consumers as there is otherwise little incentive, time, and funding for such activity. In this respect, Chapter 9 and 10's findings mirror Princen and Rhinard's research that high politics trumps low politics (2006:1122).

Decision-making and the Public Interest

As outlined in Chapter 2, the normative socio-legal literature on Internet governance has argued for ways in which to promote recognition of certain Internet rights within SDOs through existing international legal agreements (Brown and Marsden, 2009; DeNardis, 2014, 2015; Cath and Floridi, 2016). Cath and Floridi have noted the 2003 UN World Summit on the Information Society's (WSIS) Declaration of Principles, the 2012 UN Human Rights Council on 'Internet Free Speech', the 2013 UN General Assembly resolution's right to privacy, and the 2014 Netmundial Initiative (NMI) (2017:452). Scholte puts forward a polycentric approach to studying Internet governance wherein 'polycentric framework, global public policy is generated through complex networks that comprise official, market, civil society and hybrid agencies' (2010:460).

Civil rights organizations such as Article 19 and the ACLU and digital rights groups such as the Centre for Democracy and Technology, ORG and the EFF have pressured SDO fora to include civil and human rights principles. The book has shown that SDOs have taken these principles on board to some extent particularly with the establishment of the IETF's Human Rights Protocol Considerations Research Group in 2015 which encourages IETF standard development to 'enable, strengthen or threaten' human rights and resulted in a number of initiatives, notably the 451 code, as covered in Chapter 9.

In parallel, civil society groups are pushing for human rights in Internet governance within international organizations, notably the United Nations, World Economic Forum (WEF), and Organisation for Economic Co-operation and Development (OECD). In 2016, the UN Human Rights Council passed a resolution confirming the 'promotion, protection, and enjoyment of human rights on the internet'.[2] With reference to Article 19 of the UDHR and the ICCPR, it states that people should have 'the same rights that people have offline . . . [and] . . . must also be protected online'. Accordingly, in 2018, the WEF has established a Community on Global Future Council on Human Rights and Technology in conjunction with the 70th Anniversary of the Universal Declaration of Human Rights.[3] In 2018, the OECD established Guidelines for Due Diligence Guidance for Responsible Business Conduct[4] as a part of its human rights theme.[5]

These are important elements in the toolbox of global level regulation and there is scope and potential traction to translate these efforts into SDO decision-making. As covered in Chapter 2, Brown and Marsden (2009) argued for 'prosumer law' and a 'holistic regulation of the Internet' with the proposal that the WTO extend its 1996 Information Technology Agreement (ITA) to include the ten principles of the Internet Governance Forum's (IGF) multistakeholder Internet Rights and Principles Coalition (IRPC). Principle 9 specifically relates to the Internet's architecture and standards but the ITA,

[2] https://www2.ohchr.org/english/bodies/hrcouncil/docs/17session/A.HRC.17.27_en.pdf.

[3] https://www.weforum.org/communities/the-future-of-human-rights.

[4] https://mneguidelines.oecd.org/OECD-Due-Diligence-Guidance-for-Responsible-Business-Conduct.pdf.

[5] https://mneguidelines.oecd.org/themes/human-rights.htm.

which was expanded in 2015 and now includes 82 states and 201 IT products, has not yet embraced ITA principles.

There is a wider set of guidance and rules pertaining to SDOs which could make reference to public interest considerations. The International Organization for Standardization (ISO) recognizes industry standards globally. Its 2010 ISO 26000 Guidance on social responsibility incorporates seven principles: accountability, transparency, ethical behaviour, respect for stakeholder interests, respect for the rule of law, respect for international norms of behaviour, and, respect for human rights. Notably, the ISO has consultative status with the UN Economic and Social Council. Its 2016–20 strategic plan does not mention human rights but promotes 'solutions that facilitate stakeholder engagement'.[6] This denotes greater consideration of a wider number of stakeholders in the ecosphere of standard development.

The WTO 1995 Agreement on Technical Barriers to Trade (TBT) includes Annex 3: the Code of Good Practice for the Preparation, Adoption and Application of Standards[7] which sets out a number of conditions. The underlying basis for the Agreement is the fostering of international trade. The WTO treats standards on a principle of reciprocity meaning that any standard from one WTO member has to be recognized by another WTO Member and standards should not be created as non-tariff barriers to trade.

The use of international standards in this process is required to be transparent. For this reason, under the WTO Agreement, standards should be approved by a recognized body responsible for establishing rules, though compliance cannot be mandated. SDOs are encouraged under the agreement not to overlap the work of existing SDOs and to defer to international standards bodies before developing their own standard, and standards should aim to maximize performance. The code applies to 'any standardizing body within the territory of a Member of the WTO'. Every SDO implementing the TBT is registered with the ISO/IEC Information Centre in Geneva. The WTO agreement stipulates that the existence of standards developed by SDOs should be made known to the Centre. The 2000 TBT Committee Decision on Principles for the Development of International Standards, Guides and Recommendations lists six principles: transparency, openness, impartiality and consensus, relevance and effectiveness, coherence, and development dimension. A number of decisions have been taken since its adoption.[8]

[6] https://www.iso.org/files/live/sites/isoorg/files/archive/pdf/en/iso_strategy_2016-2020.pdf.
[7] https://www.wto.org/english/docs_e/legal_e/17-tbt_e.htm#annexIII.
[8] https://docs.wto.org/dol2fe/Pages/SS/directdoc.aspx?filename=q:/G/TBT/1R12.pdf.

The W3C,[9] IETF, and IEEE[10] adhere to the TBT particularly with a view to provisions on increasing interoperability and removing obstacles to trade. However, they disassociate their remit from jurisdiction. For example, the W3C states in its adherence document 'we're not on the territory of one WTO Member (but in several, but also in non-member (states), through our offices/membership open to all)'. The document also notes that the W3C is not an international body. It confirms that W3C is compatible with the WTO/TBT Code of Good Practice but that there is no specific reference to human rights as such.

Similarly, the IETF adherence document does not refer to human rights but does state that it is 'the world's largest technical professional organization dedicated to advancing technology for the benefit of humanity'. Hence, the protection of human rights has not been formally introduced to the IETF but activity can be observed via the HRPC initiated by the IRTF and presentations on human rights within plenary meetings.

In contrast to the W3C and IETF, a number of IEEE initiatives have considered human rights on a more formal basis, particularly in relation to IoT. The first principle of the IEEE's 2015 Global Initiative on Ethics of Autonomous and Intelligent Systems states that 'A/IS shall be created and operated to respect, promote, and protect internationally recognized human rights'. Its 2018 Blockchain eLearning series, 'Blockchain Governance and Human Rights', investigates how the 'potential impacts of blockchain on the human rights of nationality and privacy'. Similarly, corporate responsibility is not formally mentioned in the adherence document but it is integrated into a number of initiatives particularly those relating to the environment.

The IETF adherence document does not include reference to either human rights or corporate responsibility. However, the HRPC group has a number of initiatives relating to human rights as mentioned above. In addition to these initiatives, notably, its Informational document to Internet Filtering[11] addresses both filtering to prevent the freedom of assembly and filtering to enforce cultural norms and prevent violence and child pornography, gambling, and illegal e-commerce. The focus of the document is on the 'human rights involved as cited in the Universal Declaration of Human Rights [UDHR] which is one of the foundational documents for HRPC' and the balance between human rights and censorship in particular. The HRPC document related specifically to the

[9] https://www.w3.org/2009/07wto-std-crit.html.
[10] http://globalpolicy.ieee.org/wp-content/uploads/2017/05/IEEE16029.pdf.
[11] https://www.ietf.org/archive/id/draft-elkins-hrpc-ifilter-00.txt.

EU President's 2018 State of the Union address[12] in which Jean-Claude Juncker announced that the European Commission is launching a proposal to remove terrorist content from the web within one hour and which could subject technology companies to hefty fines if they fail to do so.

National Legal Environments

The source for inclusion of human rights or public interest considerations more generally could be global or regionally determined within the WTO or European Union respectively but these global rules must be ultimately implemented nationally. Global SDOs struggle with this stipulation as most have regional branches. Each is regulated under a different jurisdictional branch in turn. This is made more complicated by the fact that each national jurisdiction has specific rules relating to SDOs under a multitude of national Acts. In the United States alone, SDO products are subject to the 1946 Trade Mark Act which restricts the entry of goods with a false designation or origin or representation, the 1993 National Cooperative Research and Production Act (NCRPA),[13] the 2004 Standards Development Organization Advancement Act (SDOAA),[14] the 2012 Leahy–Smith America Invents Act, and the activities of the American National Standards Institute (ANSI),[15] as well as international rules, such as those stipulated under the WTO's TBT Committee.

Most of this legislation relates to antitrust law as agreements between private sector actors could be collusive. In the United States, research and development was alleviated from competition law under the 1980 Justice Department Antitrust Guidance Concerning Research Joint Ventures. As Lundqvist explains, this created a 'safe harbour' for research collaboration (2014:150). The framework was put into action under the NCRA in 1984, and updated in 1993 via the National Cooperative Research and Production Act (NCRPA).[16] The NCRPA exempted research and development markets from competition law. This provision was adopted into the 2004 SDOAA[17] which extended

[12] https://ec.europa.eu/commission/priorities/state-union-speeches/state-union-2018_en.
[13] https://www.justice.gov/archive/atr/public/press_releases/2004/204345.htm.
[14] https://www.justice.gov/archive/atr/public/press_releases/2004/204345.htm.
[15] For a list of SDOs accredited by the American National Standards Institute, see: https://www.ansi.org/standards_activities/domestic_programs/accreditation_as_developer/asd.
[16] https://www.ftc.gov/enforcement/statutes/national-cooperative-research-and-production-act-1993.
[17] https://www.justice.gov/archive/atr/public/press_releases/2004/204345.htm.

competition exemptions to SDOs. The 2004 Act cites R Hewitt Pate, Assistant Attorney General in charge of the Antitrust Division, who asserted that:

> Standards development organizations develop technical standards that are essential to the efficient functioning of our national economy. Congress has determined that the threat of treble damages pressures SDOs to restrict their standards development activities at a great cost to the United States. The Standards Development Organization Advancement Act of 2004 relieves SDOs from certain antitrust concerns and facilitates the development of pro-competitive standards.

In order to be exempted from antitrust law under the 1993 NCRPA, SDOs need to be registered with the Antitrust Division of the Department of Justice (the 'Division') and the FTC. The Division then publishes a 'notice' that the SDO has been registered in the Federal Register. Two types of entities are able to register their status under the NCRPA: joint ventures and SDOs. The Department of Justice defines an SDO as the following:

> For purposes of the Act, a SDO is a domestic or international organization that plans, develops, establishes or coordinates voluntary consensus standards using procedures that incorporate the attributes of openness, balance of interests, due process, an appeals process, and consensus in a manner consistent with the Office of Management and Budget Circular Number A-119 . . . A standards development organization does not include the parties participating in the standards development organization.[18]

There are two types of notification: original, when the SDO is formed, and supplementary, when an SDO changes its activities. The required information provided to the Division is confidential due to market sensitivity. However, information provided to the Federal Register is public.[19] SDOs must provide a letter with the name, main headquarter location of the SDO, and the nature of the SDO's activities. The main headquarter location of course determines the jurisdiction of regulation of an SDO. As mentioned in Chapter 11, the Department of Justice has different rules for each registered SDO.

SDOs are also governed under the 1995 National Technology Transfer and Advancement Act and National Standards Strategy (NSS). These mainly

[18] https://www.justice.gov/atr/filing-notification-under-ncrpa#entities.
[19] http://www.govinfo.gov.

relate to the encouragement of public procurement. Rather than developing standards in-house, government and state agencies adopt private organization standards which is seen to reduce costs and increase quality and safety. The development and use of voluntary consensus standards by government agencies is outlined by the Office of Management and Budget's (OMB) Circular A-119,[20] last revised in 2016. The circular is implemented in turn by each government and federal agency.[21]

According to the OMB Circular A-119 a '[v]oluntary consensus standards body; is defined as "a type of association, organization, or technical society that plans, develops, establishes, or coordinates voluntary consensus standards using a voluntary consensus standards development process"'. The 2016 OMB Circular A-119 dropped the 2004 SDOAA reference to 'domestic or international' in application to public procurement as 'the Federal government does not make a distinction between standards bodies based on where they are domiciled' (see later). An SDO must include the following attributes to be exempted from antitrust law:

(i) Openness: The procedures or processes used are open to interested parties. Such parties are provided meaningful opportunities to participate in standards development on a non-discriminatory basis. The procedures or processes for participating in standards development and for developing the standard are transparent.
(ii) Balance: The standards development process should be balanced. Specifically, there should be meaningful involvement from a broad range of parties, with no single interest dominating the decision-making.
(iii) Due process: Due process shall include documented and publically available policies and procedures, adequate notice of meetings and standards development, sufficient time to review drafts and prepare views and objections, access to views and objections of other participants, and a fair and impartial process for resolving conflicting views.
(iv) Appeals process: An appeals process shall be available for the impartial handling of procedural appeals.
(v) Consensus: Consensus is defined as general agreement, but not necessarily unanimity. During the development of consensus, comments and objections are considered using fair, impartial, open, and transparent processes.

[20] https://www.nist.gov/sites/default/files/revised_circular_a-119_as_of_01-22-2016.pdf.
[21] For NIST implementation, see here: https://www.nist.gov/sites/default/files/revised_circular_a-119_as_of_01-22-2016.pdf.

In this respect, the Circular confirms compliance with the WTO Agreement on Technical Barriers to Trade and other trade agreements. It also incorporates recognition of four presidential executive orders which include public interest considerations. Notably, Executive Order 12866 (Regulatory Planning and Review) states that 'the approach selected should maximize net benefits to society'; Executive Order 13563 (Improving Regulation and Regulatory Review) 'must protect public health, welfare, safety, and [the] environment'; and Executive Order 13609 (Promoting International Regulatory Cooperation) instructs Federal agencies to promote 'good regulatory practices both in the United States and internationally'. It should be noted that these provisions relate only to use of standards within national agencies, not the private sector.

ANSI also accredits standards. It keeps records of standard setting within the United States. ANSI lists Essential Requirements which refer to due process requirements for American National Standards.[22] These are openness, lack of dominance, balance, coordination and harmonization, notification of standards development, consideration of views and objections, and consensus vote. The National Institute of Standard and Technology (NIST) also has an important role to play in the promotion and funding of standards. In 1995, NIST and ANSI signed a memorandum of understanding which was last updated in 2019.[23] By 2019, ANSI had accredited over 220 SDOs.[24]

European Union

The European Union became involved in standards regulation from 1992 when it recognized ETSI as a European standards organization. This recognition was flanked by the 1992 Communication on Intellectual Property Rights and standardization. The European Commission formally recognized two further standards bodies, CEN and CENELEC which work towards harmonisation and deterring divergence in national standards. These standards bodies operate in a similar way to the W3C and IETF in that they create technical standards with representatives from industry. However, a key difference is that experts are appointed by national standards organizations rather than

[22] https://share.ansi.org/Shared%20Documents/Standards%20Activities/American%20National%20Standards/Procedures,%20Guides,%20and%20Forms/2019_ANSI_Essential_Requirements.pdf.

[23] https://share.ansi.org/shared%20documents/About%20ANSI/Memoranda%20of%20Understanding/ansinist_mou.pdf last accessed May 21, 2019.

[24] https://www.ansi.org/about_ansi/faqs/faqs.

by individual membership or appointment by companies and participant organizations. Also similar to the W3C and IETF, CEN and CENELEC agree on a consensus basis. However, in ETSI, voting is weighted by organization size. This contrasts to many SDOs, notably the W3C, where committees have one vote per organization. The European Commission only utilises standards from formally recognised European standards organisations for the purposes of public procurement. As explained by an interviewee from ISOC (interview date 30.03.2017):

> We want the European Commission to recognise IETF work and to have a seat on its multistakeholder platform. The EU only recognises formal standards e.g. from ISOs which they participate in. They don't recognise non formal standards e.g. from the IETF, W3C, IEEE. For this reason, these standards cannot be used for EU public procurement purposes. The Commission attempted to get recognition. However, there is a process under which governments need to say that IETF specification is important criteria assessment under the WTO. Namely a non-technical assessment which includes: is it an open standard, was there due process, has there been implementation, etc. The W3C does not have this problem because it has formal recognition and relations with ISO. So the W3C doesn't have to go through the WTO hurdle.

In 2010, the European Commission introduced its Horizontal cooperation agreements.[25] These provide a safe harbour for SDO agreements similar to the 2004 SDOAA in the United States (Lundqvist, 2014:199–201). In 2011, the European Commission conducted a review of its SDO policies the key purpose of which was to produce a Regulation to accredit self-regulatory standards fora. Notably the original proposal plan was for the European Commission to approve standards developed in SDOs (such as the W3C and IETF) as officially recognized European standards. However, this would have required an extra step for such fora to register additionally with the European Union once they had been registered with the WTO/TBT. The review also recommended that the European Commission could make formal proposals to external SDOs.

Following on from the review, the Commission released an impact assessment on its 'standardisation package' with a request for information from SDOs in 2011. The reaction to this from the standards community was generally

[25] Commission Communication on Guidelines on the applicability of Article 101 of the Treaty on the Functioning of the European Union to horizontal co-operation agreements (Text with EEA relevance) (2011/C 11/01).

negative. The IAB, submitting a reply on behalf of the IETF, summarizes this reaction well. It stated that 'the IAB believes there will be a lot of unnecessary overhead if IETF standards would be transposed into ESO documents.'[26] It further argued that as RFCs are often updated, there would be technical confusion and reference to one document (the RFC) rather than a possibility of two (RFC or European Standards Organization (ESO)) would also create confusion.

In response, the EU drew back from formal recognition and recognized instead the role of self-regulation in its 2012 Regulation on European Standardisation.[27] The key purpose of the Regulation was to accredit existing SDOs. The focus of the document was that standards should be voluntary and determined by the market to meet commercial operators. The Regulation incorporates the WTO principles of openness, consensus, balance, and transparency. Notably, it stated that notwithstanding this priority, standard development 'should take into account the public interest'. The 2012 Regulation importantly provides a definition for standards as 'voluntary technical specifications for products, production process, or services'.

The Commission followed this with a Joint Initiative on Standardisation in 2016[28] and a 2017 Communication on ICT Standardisation Priorities for the Digital Single Market. The latter focused on the IoT, 5G networks, and cybersecurity. On IoT, the Commission aimed to intercept interoperability failure with legal measures (presumably competition law), promote an 'open system for object identification and authentication', and to introduce guidelines for trust, privacy and security through a 'trusted IoT label'. On 5G, it is promoting the EU's lead in global industry standards via public–private partnership. On cybersecurity, it emphasized the protection of personal data. From 2017 onwards, ESOs were subject to the European Commission's regulatory fitness and performance (REFIT), namely impact assessments particularly in view of a reduction in duplication of work. Its 2018 report on EU standardization policy reiterates the Commission's promotion of the EU's role in global standard setting highlighting recent integration of standards agreement within free trade agreements (FTAs) with Canada, Singapore, Vietnam, and Japan.[29]

Although the European Commission has addressed the public interest in relation to standard setting, none of its documents have yet addressed human

[26] https://www.iab.org/wp-content/IAB-uploads/2011/03/2010-02-05-IAB-Response-Euro-ICT-Questionnaire.pdf.

[27] https://eur-lex.europa.eu/LexUriServ/LexUriServ.do?uri=OJ:L:2012:316:0012:0033:EN:PDF.

[28] http://ec.europa.eu/DocsRoom/documents/22722/attachments/1/translations/en/renditions/native.

[29] https://eur-lex.europa.eu/legal-content/EN/TXT/?uri=COM:2018:26:FIN.

rights. However, in 2017, the Economic and Social Committee (ESC) called for a code of ethics on artificial intelligence in its Opinion on Artificial intelligence: the consequences of artificial intelligence on the (digital) single market, production, consumption, employment and society. The Opinion states that 'AI systems remain compatible with the principles of human dignity, integrity, freedom, privacy and cultural and gender diversity, as well as with fundamental human rights'. In particular, it states that the European Union can maintain advantage on the global market by developing 'responsible European AI systems' including an EU AI certification and labelling system.

Transnational Cooperation Policy Convergence

There is some evidence of transnational cooperation in approaches to SDO oversight. This was shown most sharply in Chapter 11 where there has been clear convergence in the approaches of national and EU legislators and courts on SDOs' IPR policies. However, there is lingering disagreement between the United States and European Union over a wide range of SDO-related issues, particularly given the fact that the Internet was developed initially in a US legal environment. The independence of the Internet Corporation for Assigned Names and Numbers (ICANN) from the US Department of Commerce has been discussed at length in previous academic assessments of Internet governance and will not be discussed here. The main issue of ICANN independence from the US government was partially solved by the Internet Assigned Numbers Authority (IANA) transition, but there are other areas of discrepancy between different legal systems which still exist today.

Recently, there have been movements towards mutual recognition of standard developing between the EU and US. Notably, in July 2018, the EU–US Joint Statement between President Juncker and President Trump following Juncker's visit to the White House stated (under track 5) 'closer cooperation on standards'.[30] In particular 'discussions on regulatory issues and standards would be held outside the context of any possible trade negotiation'. In relation to Cybersecurity, DG Connect and NIST agreed to 'intensify their cooperation' and in particular 'standards and technical specifications developed by US-domiciled standards development organizations, may be taken into consideration in the future development of standards and voluntary certification schemes in the EU'. It also proposes certification conformity for connected

[30] http://trade.ec.europa.eu/doclib/docs/2019/january/tradoc_157651.pdf.

devices and 'cooperation on the development and use of new standards in support of regulations, for example in the area of new technologies'. The EU suggested cooperation in particular in the sectors of additive manufacturing, robotics, and automated and connected vehicles.

This was followed by a statement from the European Commission[31] encouraging 'close dialogue on standards in order to ease trade, reduce bureaucratic obstacles and slash costs'. In January 2019, the European Commission submitted a proposal to the Council for a launch of trade negotiations[32] which included a proposal for a Directive on conformity assessment with the view to removing non-tariff barriers. Here the Commission envisions the harmonization of technical standards on a transatlantic basis. Certainly there is scope for the inclusion of reference to human rights issues and corporate responsibility.

There is also evidence of a convergence in approach between the United States and the European Union on consumer privacy. The United States has historically adopted a self-regulated 'codes of conduct' approach (e.g. 'opt-out' of data tracking rather than 'opt-in') compared to other states. Individual US states have their own privacy laws, such as the 2018 California Consumer Privacy Act (CCPA). However, there has been a movement for a national Consumer Privacy Bill of Rights (Kaye, 2017). This began in 2009 in the United States with the organization of a number of roundtable discussions by the FTC on Exploring Privacy followed by publication of a paper on Commercial Data Privacy and Innovation in the Internet Economy: A Dynamic Policy Framework by the Department of Commerce in 2010.[33] The Obama Administration published a 2012 paper entitled Consumer Data Privacy in a Networked World: A Framework for Protecting Privacy and Promoting Innovation in the Global Digital Economy. This introduced the idea for a privacy 'bill of rights'. It led in 2015 to a draft bill of the Consumer Privacy Bill of Rights Act. This was not taken forward by the Trump Administration. However, in 2018, Democrats put forward proposals for a CONSENT Act which would contain a privacy 'bill of rights' in the wake of the Cambridge Analytica scandal.

Although this bill has not yet come to fruition, it has been substantially discussed in Congress and by various agencies. In September 2018, the NTIA launched a consultation on a high-level framework for protecting consumer data privacy. David Redl, Assistant Secretary for Communications and Information and National Telecommunications and Information

[31] http://europa.eu/rapid/press-release_STATEMENT-18-4687_en.htm.
[32] http://trade.ec.europa.eu/doclib/press/index.cfm?id=1971.
[33] https://www.ntia.doc.gov/files/ntia/publications/iptf_privacy_greenpaper_12162010.pdf.

Administration (NTIA) Administrator, stated: 'the Trump Administration is committed to providing real results for consumers, while giving organizations legal clarity, transparency and flexibility to innovate. We cast a wide net in asking Americans to tell us their views on privacy, and met with over 100 organizations leading up to our comment outreach.'[34] In 2019, the US Chamber of Commerce published a model for a federal law on privacy. It recommends implementation by the FTC.[35] Importantly it proposes an opt-out rather than opt-in model to consumer data sharing along with a right to deletion. Different organizations have put forth proposals, notably Intel[36] and the Centre for Democracy and Technology.[37]

In this respect, many of the more recent US proposals resemble provisions within the EU's 2016 General Data Protection Regulation (GDPR), and the proposed ePrivacy Regulation (ePR) Council of the European Union (2016). The GDPR updated the 1995 Directive on the protection of individuals with regard to the processing of personal data and on the free movement of such data. The 1995 Directive was designed for computerization/databases but not for the Internet which the GDPR rectified. Notably, it was based on fundamental rights recognized in the constitution and laws of the Member States and in the European Convention for the Protection of Human Rights and Fundamental Freedoms

Under the GDPR, personal data needs to be processed lawfully and fairly and consent needs to be a 'clear affirmative action'. This data should only be obtained for specific purposes and not be processed in any way other than for these purposes. Personal data need not only to be relevant but also be accurate and kept up-to-date and not for a longer time than it is necessary for its purpose of processing ('data retention') and can only be processed in accordance with the rights of the data subject. There is also a data export rule that personal data can only be exported outside the European Union to countries with adequate privacy protection. For this reason, states around the globe have adopted privacy legislation for compatibility with the EU's GDPR including Argentina, Brazil, Israel, and Japan. Other states such as Canada have updated existing legislation accordingly. This increases the pressure for global convergence on approaches to Internet privacy.

[34] https://www.ntia.doc.gov/press-release/2018/ntia-releases-comments-proposed-approach-protecting-consumer-privacy.

[35] https://www.uschamber.com/sites/default/files/uscc_dataprivacymodellegislation.pdf.

[36] https://usprivacybill.intel.com/legislation/.

[37] https://cdt.org/files/2018/12/2018-12-12-CDT-Privacy-Discussion-Draft-Final.pdf.

A salient issue under GDPR discussion was whether IP addresses would be considered as 'personal data' or not. Most stakeholder submissions, including the majority of EU Member States, considered IP addresses to be 'personal data' which therefore could not be passed to third parties without consent (or other ground for lawful processing). This premise was agreed upon by Member States in the Article 29 Working Party which stated in a 2007 Opinion that

> unless the Internet Service Provider is in a position to distinguish with absolute certainty that the data correspond to users that cannot be identified, it will have to treat all IP information as personal data, to be on the safe side. These considerations will apply equally to search engine operators.[38]

Significantly, this approach was formally confirmed in the 2016 *Breyer v Germany* case in the ECJ.

The GDPR obliges companies and organizations to maintain detailed internal records ('accountability') and to employ a chief privacy officer (there is an exemption for small and medium-size enterprises). Subject access requests include the right to knowledge about storage periods, rights to rectification and to erasure, and rights to lodge complaints. Information as to how data protection authorities must handle complaints is outlined in the Regulation. Breach notification was agreed at twenty-four hours (Article 67). Data protection impact assessments need to be carried out by public authorities (Article 73) and other private controllers where there is a 'high risk'.

Anonymized data remains outside the scope of the GDPR but should be covered in the proposed e-Privacy Regulation (Regulation on Privacy and Electronic Communications) (ePR) (Edwards, 2019, European Commission, 2015b). The Regulation is expected to affect direct marketing, advertising, big data, telemarketing, and recent business-to-business data sharing (B2B). The 2016 proposal was informed by a number of European Commission studies including a review of the regulatory framework for electronic communications,[39] decisions of the Article 29 Data Working Party on smartphones,[40] behavioural advertising,[41] and the desire to ensure compatibility with the GDPR and ECJ decisions. There is no mention of standards bodies or privacy considerations in

[38] https://www.clinicalstudydatarequest.com/Documents/Privacy-European-guidance.pdf.

[39] https://ec.europa.eu/digital-single-market/en/news/public-consultation-evaluation-and-review-regulatory-framework-electronic-communications.

[40] http://ec.europa.eu/justice/data-protection/article-29/documentation/opinion-recommendation/files/2013/wp202_en.pdf.

[41] http://ec.europa.eu/justice/policies/privacy/docs/wpdocs/2010/wp171_en.pdf.

standard setting within the proposal. However, there is ample scope for recognition of and implementation of both the GDPR and the ePrivacy Regulation within standard-setting for a (European Commission 2015a, 2016).

The proposed Regulation[42] aims to harmonize rules across the EU on data processing. It covers hardware and data services. It includes periods for notification of data breach, confidentiality of communications, rules on data traffic and location, unsolicited online marketing, bill itemization, call restrictions, subscriber directories, and automatic call forwarding. The European Commission proposed that the GDPR's provisions related to confidentiality should be extended to Over-The-Top (OTT) services such as unmanaged Voice over IP (VoIP), web mail, instant messaging, and instant messaging via social networks. In consultations, the Commission claimed that extension to OTTs was supported by '76% of citizens and civil society and 93.1% of public authorities agree, while only 36.2% of respondents from industry favour such an extension' **(European Commission, 2009, 2010, 2012, 2016b)**.

In relation to data tracking, the ePR stipulates transparency and consent in the tracking of cookies but also that the request for cookie consent is not repetitive. Cookies can be utilized for a single session only. In addition, the proposal stipulated that device fingerprinting be included in consent requirements.

Another key issue is the status of interception of MAC addresses. Many devices have unique identifiers, such as the MAC address of a smartphone. Trackers encrypt MAC addresses in order to utilize data but this use is a legal grey area. Examples of such tracking include tracking waiting times at airports through users' mobile phone movements (Coxworth, 2015, European Commission, 2015a). Digital rights lobbyists and some Member States view this as a legal breach of confidentiality and that the interception of data via a WiFi network adapter allows for movement tracking and ultimately recognition of the user. The ePR proposes that 'providers engaged in such practices should display prominent notices located on the edge of the area of coverage informing end-users prior to entering the defined area that the technology is in operation within a given perimeter'.

Both the United States and the European Union are showing a willingness to challenge the social media giants whose business models depend on profiling. This has also come in legal challenges at EU and national levels as well as legislative Acts and proposals. Self-regulation is expected to increase with the growth of digital technology, IoT, and the increase in cross-border data transfer on an international scale. The move towards transnational decision-making

[42] https://ec.europa.eu/newsroom/dae/document.cfm?doc_id=41241.

donates a greater reliance on expertise and technical knowledge (Ronie, 2000). The implications and scope for translation of these pieces of legislation into SDOs activity cannot be underestimated making SDOs' role and place in society increasingly central.

Conclusion

The book's findings have significance for the literatures on global governance and transnational activism by showing that spaces for civil society and such transnational activism are highly constrained in global standard-setting fora. The book confirms similar findings in the global governance literature where the public interest is marginalized in environments where the national layer of rules, laws, and policies which normally mediate interaction between citizens, corporations, and states are removed (Büthe and Mattli, 2011). However, the book adds to the literature by demonstrating that a shift in arena for decision-making to standards-developing organizations (SDOs) can present significant impasse even for well established, historically successful digital rights groups such as the Electronic Frontier Foundation.

It is important for future research to understand why actors shift arenas and which causal mechanisms enable them to do so. Evidence from the book's case studies indicates that corporate actors are attracted to SDOs away from the state. SDO decision-making is more efficient for certain actors as it changes who coordinates this decision-making. Shifting arenas stack the deck in that they filter types of influence and the resources available to them. Private self-regulatory arenas tend to privilege a certain type of values, norms, and behaviours over others. In doing so, they have causal influence on decisions. For example, Chapters 4 and 9 showed that states are able to prioritize security over consumer rights via technical standard setting. Chapter 10 demonstrated that the forum shift to SDOs for Digital Rights Management (DRM) increased the power of rights holders who gained new allies in their historical battle with digital rights activists.

Shifting areas also creates inertia. When a group enters the policy game in one arena, it is costly to shift to a different area. The political costs and technological limitations to switch to SDO fora often proved too high for digital rights activists but a lucrative investment for companies. Corporate and state actors have learned from the reframing of national debates in technical terms and are able to reinstate progress within SDO fora even though debates had been negatively framed by digital rights activists at national and international levels. This reframing is being greatly assisted by technological advances.

Therefore, some of the findings of this book have consequences for the literature on transnational activism as they illustrate how tactics utilized by digital rights groups in other global governance areas were ineffective within the unstructured environments of SDOs. This is despite the increase in transgovernmental networks as demonstrated by the literature on transnationalism (Cerny, 1995, 2010; Risse-Kappen, 1995; Strange, 1996; Scholte, 2000, 2017). The forum shift weakened the capacity for transnational activism as it removed the capacity for politicization. In particular, the lack of formal state involvement in SDOs renders decision-making less visible and vulnerable to public pressure. The move to SDOs coupled with the cross-border and transnational nature of the Internet has led to an erosion of public interest representation.

According to the global governance literature, the involvement of civil society organizations (in this case represented by consumer and digital rights groups, open source groups, and academics) might have promoted norm diffusion of new ideas and models of participatory governance. This could have triggered a change within the SDOs to greater inclusion of civil society groups in their decision-making processes, which could provide greater legitimacy and validity for the SDOs. However, to date, there is no evidence of 'complex multilateralism' or reform of SDO decision-making to accommodate civil society input. Instead, path dependency continues where wider legitimacy is constrained by institutional ambiguity, lack of public consultation over decision-making, and increasingly complex technological advances which make translation to citizens' lived experiences difficult.

Indeed, when a standard is adopted by SDOs, a form of legitimization takes place to allow its validation by a wider global community. It is likely to be more widely adopted and recognized by a wide range of industries which simplifies interoperability. The move to SDOs legitimizes debates on a global scale. This raises questions for future research. How can alternative and new routes open up SDO decision-making to wider participation, given its fundamental societal significance, in an area which, as shown by this book, is still very much dominated by large corporations?

References

Abbott, KW and Snidal, D (2009) 'The governance triangle: Regulatory standards institutions and the shadow of the state' in W Mattli and N Woods (eds), *The Politics of Global Regulation.* Princeton, NJ: Princeton University Press, pages 44–88.

Ackrill, R, Kay, A, and Zahariadis, N (2013) 'Ambiguity, multiple streams and EU policy' *Journal of European Public Policy* 20(6): 871–87.

Alvestrand, H and Lie, HW (2009) 'Development of core Internet standards: The work of IETF and W3C' in IA Bygrave and J Bing (eds), *Internet Governance: Infrastructure and Institutions.* Oxford: Oxford University Press, pages 126–46.Armstrong, D, Bello, V, Gilson, J, and Spini, D (2010) *Civil Society and International Governance: The Role of Non-State Actors in Global and Regional Regulatory Frameworks.* London: Routledge.

Arnbak, A, Asghari, H, van Eeten, M, and van Eijk, N (2014) 'Security collapse in the HTTPS market' *Communications of the Association for Computer Machinery (ACM)* 57 (10): pages 47–55.

Article 29 Working Party (2007) Opinion 4/2007 on the concept of personal data <http://ec.europa.eu/justice/policies/privacy/docs/wpdocs/2007/wp136_en.pdf>.

Article 29 Data Protection Working Party (2010) Opinion 2/2010 on behavioural advertising, 22 June <http://ec.europa.eu/justice/policies/privacy/docs/wpdocs/2010/wp171_en.pdf>.

Article 29 Working Party (2013) Advice paper on essential elements of a definition and a provision on profiling within the EU General Data Protection Regulation, 13 May <http://ec.europa.eu/justice/data-protection/article-29/documentation/other-document/files/2013/20130513_advice-paper-on-profiling_en.pdf>.

Article 29 Data Protection Working Party (2013) Opinion 02/2013 on apps on smart devices, 27 February <http://ec.europa.eu/justice/data-protection/article-29/documentation/opinion-recommendation/files/2013/wp202_en.pdf>.

Article 29 Working Party (2014) 'Comments in response to W3C's public consultation on the W3C last call working draft, 24 April 2014, Tracking preference expression (DNT)' 6 June 2014 <http://ec.europa.eu/justice/data-protection/article-29/documentation/other-document/files/2014/20140606_wp29_ts_standardisation_letter_to_w3c.pdf>.

Babaei, A, Andreoli-Fang, J, Pang, Y, and Hamzeh, B (2015) 'On the impact of LTE-U on Wi-Fi performance' *International Journal of Wireless Information Networks* 22(4): 336–44.

Bache, I (2012) 'Measuring quality of life for public policy: an idea whose time has come? Agenda setting dynamics in the European Union' *Journal of European Public Policy* 30(1): 21–38.

Ball, J (2013) 'NSA's Prism surveillance program: how it works and what it can do' *Guardian* June <http://www.theguardian.com/world/2013/jun/08/ nsa-prism-server-collection-facebook-google>.

Baumgartner, F and Jones, B (2009) *Agendas and Instability in American Politics*, 2nd edn. Chicago, IL: University of Chicago Press.

BBC (2013) 'BBC response to Ofcom's consultation "TV White Spaces: approach to coexistence"'.

Bekkers, R (2013) 'IPR Politics and practices of a representative group of standards-setting organizations worldwide'. Commissioned by the Committee on Intellectual Property Management in Standard-Setting Processes. Washington, DC: National Research Council <http://sites.nationalacademies.org/PGA/step/PGA_058712>.

Béland, D (2007) 'Ideas and institutional change in social security: conversion, layering, and policy drift' *Social Science Quarterly* 88(1): 20–38.

Benkler, Y (2006) *The Wealth of Networks*. New Haven, CT: Yale University Press.

Benkler, Y (2016) 'Degrees of freedom, dimensions of power' *Daedalus* 145(1) Winter: 18–32.

Besen, SM and Sadowsky, G (2016) 'The economics of Internet standards' in *Handbook on the Economics of the Internet*. Edited by J. M. Bauer and M. Latzer Cheltenham: Edward Elgar Publishing, pp. 211–28.

Bigo, Dr (2019) 'Beyond national security, the emergence of a digital reason of state(s) led by transnational guilds of sensitive information: the case of the Five Eyes Plus network' in B Wagner (ed.), *Research Handbook on Human Rights and Digital Technology: Global Politics, Law and International Rights*. Cheltenham: Edward Elgar, pp. 33–52.

Bing, J (2009) 'Building cyberspace: a brief history of Internet' in LA Bygrave and J Bing (eds), *Internet Governance: Infrastructure and Institutions*. Oxford: Oxford University Press, pp 8 - 47.

Black, J (2005) 'The impossibility of technology-based DRM and a modest suggestion' *Journal on Telecommunications and High Technology Law* 387(2) Spring.

Bradner, S and Contreras, J (2017) 'At Long Last, A Revised Patent Policy for IETF: What's Behind BCP79bis?' IETF News. 14 July, 2017.

Brady, HE and Collier, D (eds) (2010) *Rethinking Social Inquiry: Diverse Tools, Shared Standards*. Lanham, MD: Rowman & Littlefield.

Braithwaite, J and Drahos, P (2000) *Global Business Regulation*. New York, NY: Cambridge University Press.

Braman, S (2017) 'Internet histories: the view from the design process' *Internet Histories* 1(1): 70–78.

<https://www.ofcom.org.uk/consultations-and-statements/category-1/white-space-coexistence>.

British Entertainment Industry Radio Group (BEIRG) (2013) TV White Spaces: approach to coexistence—Response' December <https://www.ofcom.org.uk/consultations-and-statements/category-1/white-space-coexistence>.

Brookman, J (2015) 'Do Not Track 2.0' *Privacy Perspectives*, 16 July 2015 <https://iapp.org/news/a/do-not-track-2-0/>.

Brown, I and Marsden, CT (2013) *Regulating Code: Good Governance and Better Regulation in the Information Age. Information Revolution and Global Politics*. Cambridge, MA: MIT Press.

Brown, I (ed.) (2013), *Research Handbook on Governance of the Internet*. Cheltenham: Edward Elgar.

Brydon, A (2013) 'UK moves towards dynamic access for TV white space spectrum', 23 September <https://www.unwiredinsight.com/2013/tv-white-space>.

BT (2013) 'BT's response to the Ofcom Consultation Document: TV White Spaces: approach to coexistence' <https://www.ofcom.org.uk/consultations-and-statements/category-1/white-space-coexistence>.

Büthe, T and Mattli, W (2011) *The New Global Rulers: The Privatization of Regulation in the World Economy*. Princeton, NJ: Princeton University Press.

Burkart, P and McCourt, T (2006) *Digital Music Wars: Ownership and Control of the Celestial Jukebox*. Lanham: Rowman & Littlefield Publishing Group.

Bygrave, LA and Bing, J (2009) 'Models of internet governance' in LA Bygrave and J Bing (eds), *Internet Governance: Infrastructure and Institutions*. Oxford: Oxford University Press.

Bygrave, L, Schiavetta, S, Thunem, H, Lange, AB, and Phillips, E (2009) 'The naming game: governance of the domain name' in LA Bygrave and J Bing (eds), *Internet Governance: Infrastructure and Institutions*. Oxford: Oxford University Press, pages 147–212.

CableFree (2017) 'The History of WiFi: 1971 to today' 18 May, https://www.cablefree.net/wireless-technology/history-of-wifi-technology/.

Cairney, P and Jones, MD (2016) 'Kingdon's multiple streams approach: what is the empirical impact of this universal theory?' *Policy Studies Journal* 44(1): 37–58.

Cairney, P and Zahariadis, N (2016) 'Multiple streams approach: a flexible metaphor presents an opportunity to operationalize agenda setting processes' in N Zahariadis (ed.), *Handbook of Public Policy Agenda-Setting*. Cheltenham: Edward Elgar Publishing, pp. 87–105.

Cairney, P (2018) 'Three habits of successful policy entrepreneurs' *Policy & Politics* 46(2): 199–215.

California Assembly Bill No. 370, CHAPTER 390, An act to amend Section 22575 of the Business and Professions Code, relating to consumers. Approved by Governor September 27, 2013. Filed with Secretary of State September 27, 2013, http://leginfo.legislature.ca.gov/faces/billNavClient.xhtml?bill_id=201320140AB370

Camri, E (2016) 'Network, standards and culture' Selected Papers of AoIR. The 17th Annual Conference of the Association of Internet Researchers. Berlin, Germany. 5–8 October. https://journals.uic.edu/ojs/index.php/spir/article/view/8746

Castells, M (1996) *The Rise of the Network Society*. Oxford: Blackwell.

Cath, C and Floridi, L (2016) 'The design of the Internet's architecture by the Internet Engineering Task Force (IETF) and human rights' *Science and Engineering Ethics* 23(2): 449–68.

Centre for Democracy and Technology (2007), 'Consumer rights and protections in the behavioural advertising sector' <https://www.cdt.org/files/privacy/20071031consumerprotectionsbehavioral.pdf>.

Cerny, P (1995) 'Globalisation and the changing logic of collective action' *International Organisation* 49: 595–625.

Cerny, PG (2010) *Rethinking World Politics A Theory of Transnational Neopluralism*. Oxford: Oxford University Press.

Checkoway, S (2017) 'TLS 1.3 in enterprise networks' 22 July 2017 <https://www.cs.uic.edu/~s/musings/tls13-enterprises/>.

Chen, Q., Yu, G., Shan, H., Maaref, A., Li, G. Y., and Huang, A. (2016) 'Cellular Meets WiFi: Traffic Offloading or Resource Sharing?,' in IEEE Transactions on Wireless Communications, 15 (5) pp. 3354–3367, doi: 10.1109/TWC.2016.2520478.

Chirgwin, R (2018) 'Google's "QUIC" TCP alternative slow to excite anyone outside Google' *The Register* 17 January 2018.

Cimpanu, CC (2018) 'HTTP-over-QUIC to be renamed HTTP/3' 12 November 2018 <https://www.zdnet.com/article/http-over-quic-to-be-renamed-http3/>.

Cheval, C, Delaune, S, and Ryan, M (2014) 'Tests for establishing security properties' in M Maffei and E Tusost (eds), *Trustworthy Global Computing: 9th International Symposium*. Rome: TGC.

Chowdhury, AR, Wang, C, He, X, Machanavajjhala, A, and Somesh, JHA (2019) 'Outis: crypto-assisted differential privacy on untrusted servers' CoRR abs/1902.07756 <https://arxiv.org/abs/1902.07756>.

Christou, G and Simpson S (2007) *The New Electronic Marketplace: European Governance Strategies in a Globalising Economy*. London: Edward Elgar.

Christou, G and Simpson S (2006) 'The Internet and public–private governance in the European Union' *Journal of Public Policy* 26(1, April 2006): 43–61

Christou, G and Simpson, S (2009) 'New governance, the Internet and country code top level domains in Europe' *Governance* 22(4, Oct 2009): 599–624.

Cohen, MD, March, JG, and Olsen, JP (1972) 'A garbage can model of organizational choice' *Administrative Science Quarterly* 17(1): 1–25.

Comino, S, Manenti, FM and Thumm, N (2015) 'Intellectual property and innovation in information and communication technology' Joint Research Centre. Available at: http://publications.jrc.ec.europa.eu/repository/bitstream/JRC97541/jrc97541.pdf.

Contrares, J (2016) 'A tale of two layers: patents, standardization, and the Internet' *Denver University Law Review* 93(4): 853–95.

Contreras, JL (2016) 'A tale of two networks: patents, standards and the Internet' *Denver University Law Review* 93(4): 853–95.

Contrares, JL (2017) 'Essentiality and standards-essential patents' in JL Contreras (ed.), *Cambridge Handbook of Technical Standardization Law—Antitrust, Competition and Patent Law*. Cambridge: Cambridge University Press.

Cordeiro, CMK, Challapali, DB, and Shankar, NS (2005) 'IEEE802.22: the first Worldwide Wireless standard based on cognitive radios' *IEEE DySPAN* November.

Council of the European Union (2016) Position of the Council at first reading with a view to the adoption of a Regulation of the European Parliament and the Council on the protection of natural persons with regard to the processing of personal data and on the free movement of such data, and repealing Directive 95/46/EC (General Data Protection Regulation) 6 April 2016 <http://eur-lex.europa.eu/legal-content/EN/TXT/?uri=consil:ST_5419_2016_INIT>.

Coxworth, B (2015) 'JFK airport monitoring passengers' phones to provide wait times' *GizMag* 19 August <http://www.gizmag.com/jfk-airport-bliptrack/39010/>.

Cutler, CA (2003) *Private Power and Global Authority: Transnational Merchant Law in the Global Political Economy*. Cambridge: Cambridge University Press.

D'Ambrosia, J (2016) '114th IEEE 802 LMSC Plenary Session Minutes (Unconfirmed) Executive Committee Opening Meeting, Revision 0' <https://mentor.ieee.org/802-ec/dcn/16/ec-16-0179-00-00EC-802-ec-nov-2016-closing-minutes.pdf>.

Delazay, Y and Garth, BG (1998) *Dealing in Virtue: International Commercial Arbitration and the Construction of a Transnational Legal Order*. Chicago, IL: University of Chicago Press.

DeMars, WE (2005) *NGOs and Transnational Networks: Wild Cards in World Politics*. London: Pluto.

DeNardis, L (2009) *Protocol Politics: The Globalization of Internet Governance*. Boston, MA: MIT Press.

DeNardis, L (2011) *Opening Standards: The Global Politics of Interoperability*. Boston, MA: MIT Press.

DeNardis, L (2014) *The Global War for Internet Governance*. New Haven, CT: Yale University Press.

DeNardis, L (2015) 'The Internet design tension between surveillance and security' *IEEE Annals of the History of Computing* 37 (2): April–June, 72–83.

Digital TV Group (2013) 'DTG response to Ofcom 'TV White Space: Approach to Coexistence' 13 December. <https://www.ofcom.org.uk/consultations-and-statements/category-1/white-space-coexistence>.

digitaluk (2013) 'Response to Ofcom Consultation 'TV white spaces: approach to co-existence' 13 December. <https://www.ofcom.org.uk/consultations-and-statements/category-1/white-space-coexistence>.

DiMaggio, P and Powell, W (1983) 'The iron cage revisited: institutional isomorphism and collective rationality in organizational fields' *American Sociological Review* 48 (2): 147–60.

Doctorow, C (2016) 'Why the future of web browsers belongs to the biggest tech firms' *Guardian* 3 May <https://www.theguardian.com/technology/2016/may/03/future-of-web-browsers-ads-firefox-chrome-safari>.

Dorwin, D, Smith, J, Watson, M, and Bateman, A (2015) 'Encrypted media extensions' <http://www.w3.org/TR/encryptedmedia/>.

Doty, N (2019) 'Fingerprinting guidance for web specification authors' <https://www.w3.org/TR/fingerprinting-guidance/>.

Doty, N and Mulligan, D (2013) 'Internet multistakeholder processes and techno-policy standards. initial reflections on privacy at the world wide web consortium' *Journal on Telecommunications and High Technology Law* 11 (1), 135–82.

Drahos, P (2010) *The Global Governance of Knowledge: Patent offices and Their Clients.* Cambridge: Cambridge University Press.

Drake, WJ and Wilson III, EJ (eds) (2008) *Governing Global Electronic Networks: International Perspectives on Policy and Power.* Cambridge, MA: MIT Press.

Drezner, DW (2004) 'The global governance of the Internet: bringing the state back in' *Political Science Quarterly* 119: 447–98.

Drezner, DW (2008) *All Politics Is Global: Explaining International Regulatory Regimes.* Princeton, NJ: Princeton University Press.

Drissel, D (2006) 'Contesting internet governance: global dissent and disparities in the management of cyberspace resources' in *WIT Transactions on Information and Communication Technologies* 36: 297–309.

Dryzek, J (2014) 'Global deliberative democracy', in J-G Morin and A Orsini (eds), *The Essential Guide to Global Environmental Governance.* Abingdon: Routledge, pp. 76–9.

Dür, A and Mateo, G (2014) 'Public opinion and interest group influence: how citizen groups derailed the Anti-Counterfeiting Trade Agreement' *Journal of European Public Policy* 21(8): 1199–217.

Dwoskin, E (2015) "Do Not Track' compromise is pitched' *Wall Street Journal* 5 August 2015.

Dynamic Spectrum Alliance (2014) 'Letter to Marlene Dortch re. GN Docket No. 12-268 (Expanding the Economic and Innovation Opportunities ofSpectrum Through Incentive Auctions); GN Docket No. 12-354 (Amendment of the Commission's Rules with Regard to Commercial Operations in the 3550-3650 MHz Band); ET Docket No. 13-49 (Revision of Part 15 of the Commission's Rules to Permit Unlicensed National Information Infrastructure (U-NII) Devices in the 5 GHz Band)', 17 March.

Dynamic Spectrum Alliannce (2015) Response to Public consultation on the Draft RSPG Opinion on Common Policy Objectives for WRC-15. <http://rspg-spectrum.eu/public-consultations/>.

Dynamic Spectrum Alliannce (2015) 'Letter: Re: ET Docket No. 15-105: Office of Engineering and Technology and Wireless Telecommunications bureau seek information on current trends in LTE-U and LAA Technology' 11 June 2015 <https://ecfsapi.fcc.gov/file/60001077791.pdf#viewer.action=download>.

Eckersley, P (2010) 'How unique is your browser?' Electronic Frontiers Foundation, pages 1–19. https://panopticlick.eff.org/static/browser-uniqueness.pdf.

EE (2014) Letter to Mark Binns (Ofcom Spectrum Policy Group) on the 2013 Ofcom Consultation on TV White Spaces coexistence 14 January. <https://www.ofcom.org.uk/consultations-and-statements/category-1/white-space-coexistence>.

Economist (2004) 'A brief history of Wi-Fi' 12 June. <https://www.economist.com/technology-quarterly/2004/06/10/a-brief-history-of-wi-fi>.

ECSIP (2014) 'Patents and standards A modern framework for IPR-based standardization' Study commissioned by the European Commission, Directorate-General for Enterprise and Industry <http://ec.europa.eu/DocsRoom/documents/4843/attachments/1/translations/en/renditions/pdf>.

Ermert, M. (2017) 'QUIC has almost arrived: The IETF introduces a new internet transport protocol', 5 April 2017.

Edwards, L (2019) 'Data Protection and e-Privacy: From Spam and Cookies to Big Data Machine Learning and Profiling.' in *Law, Policy and the Internet*. Edited by L. Edwards. Oxford: Hart Publishing, pages 119–164.

Englehardt, S and Narayanan Princeton, A (2016) 'Online Tracking: a 1-million-site measurement and analysis' ACM CCS 2016 <http://randomwalker.info/publications/OpenWPM_1_million_site_tracking_measurement.pdf>.

Ericsson, Smith G, and InterDigital (2017) 'In the matter of the Appeal of Ericsson, Graham Smith and InterDigital Concerning Action Taken by the IEEE-SA Standards Board at the close of the Investigation into the 802.11ax Dominance Complaint' 5 January, http://ieee802.org/appeal_decisions/Ericsson_Smith_InterDigital_17_0106/Appeal_Brief_and_Appendix_SASB_Appeal_(2017.01.05).pdf

European Union (2002) Directive 2002/58/EC of the European Parliament and of the Council of 12 July 2002 concerning the processing of personal data and the protection of privacy in the electronic communications sector (Directive on privacy and electronic communications). *OJLL* 201 31 July 2002: *0037–0047* <http://eur-lex.europa.eu/LexUriServ/LexUriServ.do?uri=CELEX:32002L0058:en:HTML>.

European Union (2009). Directive 2009/136/EC of the European Parliament and of the Council of 25 November 2009 amending Directive 2002/22/EC on universal service and users' rights relating to electronic communications networks and services, Directive 2002/58/EC concerning the processing of personal data and the protection of privacy in the electronic communications sector and Regulation (EC) No 2006/2004 on co-operation between national authorities responsible for the enforcement of consumer protection laws.

European Commission (2009) Consultation on the legal framework for the fundamental right to protection of personal data, 31 December 2009 <http://ec.europa.eu/justice/newsroom/data-protection/opinion/090709_en.htm>.

European Commission (2010) Communication from the Commission to the European Parliament, the Council, the Economic and Social Committee and the Committee of the Regions: a comprehensive approach on personal data protection in the European Union Brussels, COM(2010) 609, 4 November 2010 <http://ec.europa.eu/justice/news/consulting_public/0006/com_2010_609_en.pdf>.

European Commission (2011) Consultation on the Commission's comprehensive approach to personal data protection in the European Union, 15 January 2011 <http://ec.europa.eu/justice/newsroom/data-protection/opinion/101104_en.htm>.

European Commission (2015a) ePrivacy Directive: assessment of transposition, effectiveness and compatibility with proposed Data Protection Regulation prepared for the EC by Time.lex and Spark, 10 June 2015. <https://ec.europa.eu/digital-single-market/en/news/eprivacy-directive-assessment-transposition-effectiveness-and-compatibility-proposed-data>.

European Commission (2015b) Public Consultation on the evaluation and the review of the regulatory framework for electronic communications networks and services, 11 September 2015 <https://ec.europa.eu/digital-single-market/en/news/public-consultation-evaluation-and-review-regulatory-framework-electronic-communications>.

European Commission (2016c) Public consultation on the evaluation and review of the ePrivacy Directive, 11 April 2016. <https://ec.europa.eu/digital-single-market/en/news/public-consultation-evaluation-and-review-eprivacy-directive>.

European Data Protection Surpvisor (2017), Annual Report, https://edps.europa.eu/sites/edp/files/publication/17-04-27_annual_report_2016_en_1.pdf.

European Data Protection Supervisor (2016), Opinion 5/2016, Preliminary EDPS Opinion on the review of the ePrivacy Directive (2002/58/EC), July 2016, https://edps.europa.eu/sites/edp/files/publication/16-07-22_opinion_eprivacy_en.pdf.

European Union (2016) General Data Protection Regulation 2016/79 of the European Parliament and of the Council of 27 April 2016, I. 119/1, 4 May 2016.

Faith Cranor, L (2012) 'Necessary but Not Sufficient: Standardized Mechanisms for Privacy Notice and Choice', http://www.jthtl.org/content/articles/V10I2/JTHTLv10i2_Cranor.PDF

Falleti, TG and Lynch, JF (2009) 'Context and causal mechanisms in political analysis' *Comparative Political Studies* 42: 1143–66.

Farrell, H (2006) 'Regulating information flows: states, private actors, and e-commerce' *Annual Review of Political Science* 9: 353–74.

Farrell, S (2017), TLS is not for obligatory Interception Lovers, https://github.com/sftcd/tinfoil.

Farrell, S (2014), Pervasive monitoring is an attack, best current practice, Internet Engineering Task Force <https://tools.ietf.org/html/rfc7258.>.

Federal Communications Commission (2002) 'Notice of Inquiry in the Matter of Additional Spectrum for Unlicensed Devices Below 900 MHz and in the 3 GHz Band' FCC 02-328, ET Docket No. 02-380, 20 December, https://docs.fcc.gov/public/attachments/FCC-02-328A1.pdf.

Federal Communications Commission (2004) 'Notice of Proposed Rulemaking in the Matter of Unlicensed Operation in the TV Broadcast Bands, Additional Spectrum for Unlicensed Devices Below 900 MHz and in the 3 GHz Band' FCC 04-113, ET Docket No. 04-186 ET Docket No. 02-380, 25 May, http://www.naic.edu/~phil/rfi/fccactions/FCC-04-113A1.pdf

Federal Communications Commission (2006) 'First Report and Order and Further Notice of Proposed Rulemaking in the Matter of Unlicensed Operation in the TV Broadcast Bands, Additional Spectrum for Unlicensed Devices Below 900 MHz and in the 3 GG\Hz Band' ET Docket No. 04-186 and ET Docket No. 02-380, FCC 06-156, 18 October, https://www.federalregister.gov/documents/2006/11/17/E6-18907/unlicensed-operation-in-the-tv-broadcast-bands

Federal Communications Commission (2008) 'Second Report and Order and Memorandum Opinion and Order in the Matter of Unlicensed Operation in the TV Broadcast Bands, Additional Spectrum for Unlicensed Devices Below 900 MHz and in the 3 GHz Band' ET Docket No. 04-186 and ET Docket No. 02-380, FCC 08-260, 14 November, https://docs.fcc.gov/public/attachments/FCC-08-260A1.pdf.

Federal Communications Commission (2010) 'Second Memorandum Opinion and Order in the Matter of Unlicensed Operation in the TV Broadcast Bands and Additonal Spectrum for Unlicensed Devices Below 900MHz and in the 3GHz Band' ET Docket

No.04-186 and ET Docket No. 02-380, FCC 10-174, 23, September, https://docs.fcc.gov/public/attachments/FCC-10-174A1.pdf.

Federal Trade Commission (2010) Prepared statement of the Federal Trade Commission on Do Not Track, 2 December. ET Docket No. 02-380, FCC 10-174, 23 September <http://ftc.gov/os/testimony/101202donottrack.pdf>.

Federal Trade Commission (2012) Protecting consumer privacy in an era of rapid change: recommendations for businesses and policymakers, 26 March 2012 <https://www.ftc.gov/reports/protecting-consumer-privacy-era-rapid-change-recommendations-businesses-policymakers>.

Federal Trade Commission (2017) Businesses can help stop phishing and protect their brands using email authentication. March <https://www.ftc.gov/system/files/documents/reports/businesses-can-help-stop-phishing-protect-their-brands-using-email-authentication-ftc-staff/email_authentication_staff_perspective.pdf>.

Feldman, A (2017) 'Encryption issues in the IETF's QUIC Standard'. Presentation at 'Current Issues in Standards Development Organisations Decision-Making Workshop', 15 November 2017, Portcullis House, London.

Filin, S, Noguet, D, Dore, J-B, Mawlawi, B, Holland, O, Shakir, MZ, Harada, H, and Kojima, F (2018) 'IEEE 1900.7 Standard for White Space dynamic spectrum access radio systems' *IEEE Communications Magazine* 56(1): 188–92.

Finnemore, M and Sikkink, K (1998) 'International norm dynamics and political change' *International Organization* 52(4): 887–917.

Froomkin, M (2013) 'ICANN and the domain name system after the "Affirmation of Commitments"' in I Brown (ed.), *Research Handbook on Governance of the Internet*. Cheltenham: Edward Elgar.

Gajek, S, Manulis, M, Pereira, O, Sadeghi, A-R, and Schwenket, J (2008) 'Universally composable security analysis of TLS' <https://eprint.iacr.org/2008/251.pdf>.

Garcia-Johnson, R. (2000) *Exporting Environmentalism: U.S. Multinational Chemical Corporations in Brazil and Mexico*. Cambridge, Mass.: MIT Press.

Gencer, M (2012) 'The evolution of IETF standards and their production' *International Journal of IT Standards and Standardization Research* 10(1, January–June): 17–33.

Gerring, J (2006) *Case Study Research: Principles and Practices*. Cambridge: Cambridge University Press.

Gerring, J (2010) 'Causal mechanisms: Yes, but . . . ' *Comparative Political Studies* 43(11): 1499–526.

Gibbs, S (2013) 'White space broadband: your questions answered' *Guardian* 2 October <https://www.theguardian.com/technology/2013/oct/02/white-space-broadband-microsoft-google-wireless-rural>.

Gilb, J (2016) 'IEEE 802.11ax Dominance Allegation' IEEE802 LMSC EC, ec-16-0197-02-00EC, November.

Gilpin, L (2014) 'White Space, the next internet disruption: 10 things to know' 12 March <https://www.techrepublic.com/article/white-space-the-next-internet-disruption-10-things-to-know/a>.

Glen, CM (2014) 'Access Internet governance: territorializing cyberspace?' *Politics & Policy* 42(5): 635–57.

Goldstein, J and Steinberg, RH (eds) (2010) *International Institutions*. London: Sage.

Goldsmith, J and Wu, T (2006) *Who Controls the Internet? Illusions of a Borderless World*. Oxford: Oxford University Press.

Google (2019) 'TV White Spaces: approach to coexistence' <https://www.ofcom.org.uk/consultations-and-statements/category-1/white-space-coexistence>.

Götz, Nt (2008) 'Reframing NGOs: the identity of an international relations non-starter' *European Journal of International Relations* 14(2): 231–58.

Grasso, R (2018) 'Selected issues in SEP licensing in Europe: the antitrust perspective' in A Bharadwaj, VH Devaiah, and I Gupta (eds), *Complications and Quandaries in the ICT Sector.* New York, NY: Springer, pp. 79–97.

Gray, LK (2017) 'Are you ready for 30 June 2018? Saying goodbye to SSL/early TLS' *eCommerce and Patching and Awareness and TLS/SSL and PCI DSS and Encryption. PCI Security Standards* <https://blog.pcisecuritystandards.org/are-you-ready-for-30-june-2018-sayin-goodbye-to-ssl-early-tls>.

Green, J (2010) 'Private authority on the rise: a century of delegation in multilateral environmental agreements' in C Jönsson ad J Tallberg (eds), *Transnational Actors in Global Governance. Patterns, Explanations and Implications.* Oxford: Oxford University Press, pages 155 - 173.

Greenstein, S (2007) 'Economic Experiments and Neutrality in Internet Access' Working Paper 13158, NBER Programs, June <http://www.nber.org/papers/w13158>.

Greenstein, S (2017) *How the Internet Became Commercial: Innovation, Privatization, and the Birth of a New Network.* Princeton, NJ: Princeton University Press.

Greenwald, Glenn (2013) 'XKeyscore: NSA tool collects "nearly everything a user does on the internet"' *Guardian* July 2013 <http://www.theguardian.com/world/2013/jul/31/nsa-top-secret-program-online-data>.

Greenwald, G (2014) *No Place to Hide: Edward Snowden, the NSA, and the U.S. Surveillance State.* London: Hamish Hamilton.

Grugel, J and Iusmen, I (2013) 'The European Commission as guardian angel: the challenges of agenda-setting for children's rights' *Journal of European Public Policy* 1(20): 77–94.

Guadamuz González, A (2006) 'The software patent debate'. *Journal of Intellectual Property Law and Practice* 1(3): 196.

Gunes, A, Juarez, M, Nikiforakis, N, Diaz, C, Gürses, S, Piessens, F, and Preneel, B (2013) 'FPDetective: dusting the web for fingerprinters' ACM Conference on Computer and Communications Security (CCS 2013), November, Berlin, pp 1129–40 <https://dl.acm.org/citation.cfm?id=2508859>.

Hacker, J (2005) 'Policy drift: the hidden politics of US welfare state retrenchment' in W Streeck and K Thelen (eds), *Beyond Continuity: Institutional Change in Advanced Political Economies.* Oxford: Oxford University Press, pp. 40–82.

Hall, RB and Biersteker, TB (2002) *The Emergence of Private Authority in Global Governance.* Cambridge: Cambridge University Press.

Halpin, H (2016) 'The responsibility of open standards in the era of surveillance. Hot topics in privacy enchancing technologies' HAL-INRIA open archive, July, Darmstadt, Germany. hal-01426848. https://hal.inria.fr/hal-01426848/document.

Abelson, H, Anderson, R, Bellovin, SM, Benaloh, J, Blaze, M, Diffie, W, Gilmore, J, Green, M, Landau, S, Neumann, PG, Rivest, RL, Schiller, JI, Schneier, B, Specter, M, and Weitzner, DJ (2015) 'Keys under doormats: mandating insecurity by requiring government access to all data and communications' *Computer Science and Artificial Intelligence Laboratory Technical Report* Massachusetts Institute of Technology. 6 July 2015 <https://dspace.mit.edu/handle/1721.1/97690>.

Haunss, S and Kohlmorgen, L (2010) 'Conflicts about intellectual property claims: the role and function of collective action networks' *Journal of European Public Policy* 17(2): 242–62.

Held, D (1995) *Democracy and the Global Order*. Palo Alto, CA: Stanford University Press.

Hengesbaugh, B and de La Lama, A (2013) 'How should I respond to California's Do-Not-Track requirements?' *The Privacy Advisor* 26 November <https://iapp.org/news/a/how-should-i-respond-to-californias-do-not-track-requirements/>.

Hofmann, J, Katzenbach, C and Gollatz, K (2017) 'Between coordination and regulation: finding the governance in Internet governance' *New Media & Society* 19(9): 1406–23.

Horten, M (2011) *The Copyright Enforcement Enigma: Internet Politics and the 'Telecoms Package*. London: Palgrave Macmillan.

Horten, M (2013) *The Copyright Masquerade*. London: Zed Books.

House of Lords Select Committee on the European Union (2014) EU Data protection Law: a 'right to be forgotten? Chapter 4: should there continue to be a 'right to be forgotten?' 23 July <http://www.publications.parliament.uk/pa/ld201415/ldselect/ldeucom/40/4006.htm>.

Housley, R and Droms, R (2017) 'TLS 1.3 Option for Negotiation of visibility in the datacenter draft-rhrd-tls-tls13-visibility-00' 29 September 2017 <https://tools.ietf.org/html/draft-rhrd-tls-tls13-visibility-00>.

Hung, M (2018) 'Leading the IoT: Gartner insights on how to lead in a connected world' Gartner Research <https://www.gartner.com/imagesrv/books/iot/iotEbook_digital.pdf>.

Huston, G (2018) 'Just one bit' The ISP Column: A Column on Things Internet, 2018 https://www.potaroo.net/ispcol/2018-03/onebit.html.

IETF (2000) RFC 2804 IETF Policy on Wiretapping. May <https://tools.ietf.org/html/rfc2804 >.

IETF (2014) RFC 7824 Pervasive monitoring is an attack. May <https://tools.ietf.org/html/rfc7258 >.

IEEE (2015) 'Follow-up liaison statement regarding LAA' 18 May 2015 <http://ieee802.org/secmail/pdfuYWRYdEDz5.pdf.

IEEE (2015) 'Liaison statement regarding LAA/802.11 Coexistence, 802 LMSC Plenary, 13 March 2015 <https://mentor.ieee.org/802-ec/dcn/15/ec-15-0026-00-00EC-3gpp-march-2015-liaison-2-final.pdf>.

IEEE (2015) 'Liaison to 3GPP, for presentation at 29-Aug-15 LAA Workshop' 17 August 2015 <http://www.ieee802.org/Communications.shtml>.

IEEE (2016) 'Review of 3GPP LAA Specification Rel. 13' 1 August 2016 <http://www.ieee802.org/Communications/16_08/802_to_3GPP_01AUG_2016_Liaison_r01.pdf>.

IEEE (2016) 'The 11ax Dominance Complaint', EC-16-0186-01-00EC, November. <https://mentor.ieee.org/802.11/documents?n=2&is_dcn=DCN%2C%20Title%2C%20Author%20or%20Affiliation&is_group=coex >.

IETF (2018) RFC 8446 The Transport Layer Security (TLS) Protocol Version 1.3, August 2018 <https://datatracker.ietf.org/doc/rfc8446/ >.

Internet Society (2019) Internet Society Global Internet Report Consolidation in the Internet Economy https://future.internetsociety.org/2019/.

Jeon, J, Niu, H, Li, Q, Papathanassiou, A, and Wu, G (2015) 'LTE with Listen-Before-Talk in unlicensed spectrum' 2015 IEEE International Conference on Communication Workshop (ICCW) <http://ieeexplore.ieee.org/stamp/stamp.jsp?arnumber=7247527>.

Jones, BD (2001) *Politics and the Architecture of Choice: Bounded Rationality and Governance*. Chicago, IL: University of Chicago Press.

Jönsson, C and Tallberg, J (eds) (2010) *Transnational Actors in Global Governance. Patterns, Explanations and Implications*. London: Palgrave MacMillan.

Jupille, J, Mattli, W, and Snidal, D (2013) *Institutional Choice and Global Commerce*. New York, NY: Cambridge University Press.

Kalm, S (2010) 'Limits to transnational participation: the global governance of migration' in CJJ Tallberg (ed.), *Transnational Actors in Global Governance*. London: Palgrave Macmillan, pp. 134–54.

Kamara, I and Kosta, E (2016), 'Do Not Track initiatives: regaining the lost user control' *International Data Privacy Law* 6(4): 276–90.

Kanevskaia, O and Zingales, N (2016) 'The IEEE-SA patent policy update under the lens of EU competition law' *European Competition Journal* 23: 1–41.

Katznelson, RD (2016) 'The IEEE controversial policy on Standard Essential Patents—the empirical record since adoption' Symposium on Antitrust, Standard Essential Patents, and the Fallacy of the Anticommons Tragedy, Berkeley, CA (29 October 2016) <http://bit.ly/IEEE-LOAs>.

Kaye, K (2017) 'New privacy report already removed from White House Report' 20 January <https://adage.com/article/privacy-and-regulation/privacy-report-removed-white-house-site/307632>.

Keohane, RO and Nye, JS (1996) 'Power and interdependence in the information age' *Foreign Affairs* 77(5): 81–94.

Khalifa, O and binti Yusof, Y (2012) 'An improved wavelet digital watermarking software implementation' International Conference on Computer and Communication Engineering (ICCCE 2012), 3–5 July 2012, Kuala Lumpur, Malaysia.

Kingdon, JW (1984) *Agendas, Alternatives and Public Policies*. Harlow: Longman.

Kingdon, JW (1995) *Agendas, Alternatives and Public Policies*, 2nd edn. Harlow: Longman.

Kingdon, JW (2011), *Agendas, Alternatives, and Public Policies*, 4th edn. London: HarperCollins.

Kingdon, JW (2013) *Agendas, Alternatives, and Public Policies*. Harlow: Pearson New International edn.

Kosta, E (2013) 'Peeking into the cookie jar: the European approach towards the regulation of cookies' *International Journal of Law and Information Technology* 21(4): 380–406.

Krawczyk, H, Paterson, K, and Wee, H (2013) 'On the security of the TLS Protocol: a systematic analysis' <https://eprint.iacr.org/2013/339.pdf>.

Kwon, H-J, Jeon, J, Bhorkar, A, Ye, Q, Harada, H, et al. (2016) 'Licensed-assisted access to unlicensed spectrum in LTE Release 13' IEEE Communications Magazine on LTE Evolution, pages 1 – 10, https://arxiv.org/pdf/1604.08632.pdf (accessed 10 October 2016).

Kühn, K-U, Morton, FS, and Shelanski, H (2013) 'Standard setting organizations can help solve the standard essential patents licensing problem' *CPI Antitrust Chronicle* 1–5.

Landau, S (2015) 'NSA and dual EC_DRBG: deja vu all over again?' *The Mathematical Intelligencer* 37(4, December): 72–83.

Lee, E (2013) *The Fight for the Future: How People Defeated Hollywood and Saved the Internet For Now*. Morrisville, NC: Lulu Press.

Lee, M (2008) 'HTML5 comes to fruition' *InformationWeek* 31 March: 48–9.

Lessig, L (2001) *The Future of Ideas*. New York, NY: Random House.

Levy, S (2001) *Crypto: How the Code Rebels Beat the Government—Saving Privacy in the Digital Age*. London: Penguin Press Science.

Liese, A (2010) 'Explaining varying degrees of openness in the Food and Agriculture Organization of the United Nations (FAO)' in CJJ Tallberg (ed.), *Transnational Actors in Global Governance*. London: Palgrave Macmillan, pp. 88–109.

Littoz-Monnet, A (ed.) (2017) *The Politics of Expertise in International Organizations. How International Bureaucracies Use and Produce Knowledge*, London: Routledge.

Loya, T and Boli, J (1999) 'Standardization in the world polity: technical rationality over power' in J Boli and G Thomas (eds), *Constructing World Culture: International Nongovernmental Organizations Since 1875*. Stanford, CA: Stanford University Press, pp. 169–97.

Lukasz O, Gunes, A, Castelluccia, C, and Diaz, C (2016) 'The leaking battery: a privacy analysis of the HTML5 Battery Status API' International Association for Cryptologic Research. IACR Cryptology ePrint Archive. <https://www.iacr.org/cryptodb/data/paper.php?pubkey=26437>.

Lundqvist, B (2014) *Standardization Under EU Competition Rules And US Antitrust Laws: The Rise and Limits of Self-Regulation*. Cheltenham: Edward Elgar.

MacAskill, E, Borger, J, Hopkins, N, Davies, N, and Ball, J (2013) 'How does GCHQ's internet surveillance work?' *Guardian* June.

Malcolm, J (2008) *Multi-Stakeholder Governance and the Internet Governance Forum*. Perth: Terminus Press.

Mallinson, K (2015) 'Cumulative mobile-SEP royalties' report by Wiseharbor consultancy group. August 19, https://www.wiseharbor.com/pdfs/Mallinson%20on%20cumulative%20mobile%20SEP%20royalties%20for%20IP%20Finance%202015Aug19.pdf

March, JG and Olsen, JP (1984) 'The new institutionalism: organizational factors in political life' *American Political Science Review* 78 (September): 734–49.

Marsden, CT (2012) 'Internet co-regulation: European law, regulatory governance and legitimacy in cyberspace' *The Cambridge Law Journal* 71(2): 457–8.

Masonta, MT, Makgotlho, R, and Mekuria, F (2012) 'Setting the scene for TV White Spaces and dynamic spectrum access in South Africa' in P Cunningham and M Cunningham (eds), IIMC Information Management Corporation, IST-Africa 2012 Conference Proceedings, pp 1–10.

Mathiason, J (2008) *The New Frontier of Global Institutions*. London: Routledge.

Mattli, W and Büthe, T (2005) 'Global private governance: lessons from a national model of setting standards in accounting' *Law and Contemporary Problems* 68: 225–62.

Mattli, W and Büthe, T (2013) *The New Global Rulers The Privatization of Regulation in the World Economy* Princeton, NJ: Princeton University Press.

Maurer, U and Tackmann, B (2010) 'On the soundness of authenticate-then-encrypt: formalizing the malleability of symmetric encryption' <https://www.researchgate.net/publication/221609593_On_the_soundness_of_authenticate-then-encrypt_Formalizing_the_malleability_of_symmetric_encryption>.

Mazzone, J (2011) *Copyfraud and Other Abuses of Intellectual Property Law*. Stanford, CA: Stanford University Press.

McKinsey Global Institute (2015) 'The Internet of Things: mapping the value beyond the hype' <https://www.mckinsey.com/~/media/McKinsey/Business%20Functions/McKinsey%20Digital/Our%20Insights/The%20Internet%20of%20Things%20The%20value%20of%20digitizing%20the%20physical%20world/The-Internet-of-things-Mapping-the-value-beyond-the-hype.ashx>.

Ménière, Y. (2015) 'Fair, Reasonable and Non-Discriminatory (FRAND) Licensing Terms' Joint Research Centre. Available at: http://publications.jrc.ec.europa.eu/repository/bitstream/JRC96258/jrc96258.pdf.

Moriarty, K (2017) 'TLS security and data centre monitoring: searching for a path forward' 20 October 2017 <https://blog.apnic.net/2017/10/20/tls-security-data-centre-monitoring-searching-path-forward/>.

Morrissey, P, Smart, N, and Warinschi, B (2008) 'A modular security analysis of the TLS Handshake Protocol' <https://eprint.iacr.org/2008/236.pdf>.

Mueck, M and Noguet, D (2011) 'TV white space standardization and regulation in Europe' <https://www.researchgate.net/publication/224244913_TV_White_Space_standardization_and_regulation_in_Europe>.

Mueller, M (2010) *Networks and States: The Global Politics of Internet Governance. Information Revolution and Global Politics.* Cambridge, MA: MIT Press.

Mueller, M, Schmidt, A, and Kuerbis, B (2013) 'Internet security and networked governance in international relations' *International Studies Review* 15(1, March): 86–104.

Musiani, F, Cogburn, DL, DeNardis, L, and Levinson, NS (2016) *The Turn to Infrastructure in Internet Governance.* New York, NY: Palgrave McMillan.

Neul Ltd (2013) 'Ofcom TV White-Space consultation: response from Neul Ltd' November <https://www.ofcom.org.uk/__data/assets/pdf_file/0032/53879/neul.pdf>.

Nominet (2018) 'What is TV White Space?' <https://www.nominet.uk/researchblog/what-is-tv-white-space/>.

Nye, J (2014) 'The regime complex for managing global cyber activities' Global Commission on Internet Governance Paper Series, 1 <http://www.cigionline.org/publications/regime-complex-managingglobal-cyber-activities>.

O'Brien, R (2006) 'GSP Forum: international relations and global social policy' *Global Social Policy* 6 (1): 14–16.

O'Brien, R, Goetz, AM, Scholte, JA, and Williams, M (2000) *Contesting Global Governance: Multilateral Economic Institutions and Global Social Movements.* Cambridge: Cambridge University Press.

OECD (2011) *Policy Round Tables. Standard Setting.* 8 March. https://www.oecd.org/daf/competition/47381304.pdf.

Ofcom (2015) 'Ofcom gives green light for "TV white space" wireless technology' 12 February <https://www.ofcom.org.uk/about-ofcom/latest/media/media-releases/2015/tvws-statement>.

O'Neill, K, Balsiger, J, and VanDeveer, SD (2004) 'Actors, norms, and impact: recent international cooperation theory and the influence of the agent-structure debate' *Annual Review of Political Science* 7: 149–75.

O'Neill, M (2018) 'Do Not Track and the GDPR', W3C Blog, https://www.w3.org/blog/2018/06/do-not-track-and-the-gdpr/.

Open Technology Institute (2015) 'Reply comments of Open Technology Institute at New America, Public Knowledge, Free Press, Common Cause In the Matter of Office of Engineering and Technology and Wireless Telecommunications Bureau Seek Information on Current Trends in LTE-U and LAA Technology' Federal Communications Commission ET Docket No 15-105, 5 May. Washington, DC 20554.

Paolini, M (2015) 'LTE unlicensed and Wi-Fi: moving beyond coexistence'. Senza Fili consultancy <https://ecfsapi.fcc.gov/file/60001076664.pdf>.

Patterson, KG and van den Merwe, T (2016) 'Reactive and Proactive Standardisation of TLS', in Chen, L, McGrew, D, and Mitchell, C. (eds) Security Standardisation Research. SSR 2016. Lecture Notes in Computer Science, vol 10074. Springer, Cham, pp: 160–186.

Pentheroudakis, C and Baron, JA (2015) 'Licensing terms of standard essential patents: a comprehensive analysis of cases' Joint Research Centre.

Pentheroudakis, C and Thumm, N (2015) 'Innovation in the Digital single market: the role of patents' Joint Research Centre of the European Commission, Brussels. Available at: <http://publications.jrc.ec.europa.eu/repository/bitstream/JRC96728/jrc96728.pdf>.

Powar, S (2017) 'Six WiFi trends that show why it's the Internet's growth business' 23 February <https://boostandco.com/wifi-trends-2017>.

Pardue, L (2019) 'HTTP/3: from root to tip' *Cloudflare blog*, 24 January <https://blog.cloudflare.com/http-3-from-root-to-tip/>.

Peacock, S (2014) 'How web tracking changes user agency in the age of Big Data: the used user' *Big Data & Society* July–December, pp. 1–11.

Perlroth, N, Larson, J and S Shane (2013) 'N.S.A. Able to foil basic safeguards of privacy on web' *New York Times* 5 September.

Perzanowski, A and Schultz, J (2017) *The End of Ownership Personal Property in the Digital Economy*. Cambridge, MA: MIT Press.

Picciotto, S (2011) *Regulating Global Corporate Capitalism*. Cambridge: Cambridge University Press.

Pilgrim, M. (2010) *HTML5: Up and Running: Dive into the Future of Web Development 1st Edition*. Sebastopol, CA: O'Reilly Media.

Porter, T and Ronet, K (2010) *The Challenges of Global Business Authority: Democratic Renewal, Stalemate, or Decay?* New York, NY: SUNY Press.

Porter, T and Ronit, K (eds) (2010) *The Challenges of Global Business Authority: Democratic Renewal, Stalemate, or Decay?* New York, NY: SUNY Press.

Post, R (2000) 'Encryption source code and the first amendment' *Berkeley Technology Law Journal* 15(2, Spring): 713–23.

Postigo, H (2010) 'Information communication technologies and framing for backfire in the digital rights movement: the case of Dmitry Sklyarov's advanced e-Book processor' *Social Science Computer Review* 28 (2): 232–50.

Princen, S (2007) 'Agenda-setting in the European Union: a theoretical exploration and agenda for research' *Journal of European Public Policy* 14: 21–38.

Princen, S and Rhinard M (2006) 'Crashing and creeping: agenda-setting dynamics in the European Union' *Journal of European Public Policy* 13(7): 1119–32.

Prakash, A and Potoski, M (2006) 'Racing to the bottom? Trade, environmental governance, and ISO 14001' *American Journal of Political Science*. 50(2, April): 350–64.

Prosser, T (1997) *Law and the Regulators*. Oxford: Clarendon Press.

Prosser, T (2010a) *The Regulatory Enterprise: Government, Regulation and Legitimacy*. Oxford: Oxford University Press.

Prosser, T. (2010b) *The Regulatory State: Constitutional Implications*. Oliver, D., Prosser, T. & Rawlings, R. (eds.). Oxford University Press, p. 34–49.

Prosser, T (2011) 'Regulation and legitimacy' in J Jowell, D Oliver, and C O'Cinneide (eds), *The Changing Constitution*. Oxford: Oxford University Press, pages 339–358.

Rashid, I and Simpson S (2019a) 'The struggle for co-existence: communication policy by private technical standards making and its limits in unlicensed spectrum' Information, Communication and Society, DOI: 10.1080/1369118X.2019.1662072.

Rashid, I and Simpson S (2019b) 'Civil society activism, strategic alignment and international public policy making for spectrum' *Journal of Civil Society* 15(1): 62–81.

Régibeau, P, De Coninck, R, and Zenger, H (2011) Transparency, Predictability, and Efficiency of SSO-based Standardization and SEP Licensing. A Report for the European Commission. Prepared for the Executive Agency for Small and medium-sized enterprises on behalf of the European Commission as part of Specific Contract No 07/2015 Implementing Framework Contract No MARKT/2011/022/B2/ST/FC.

Rimmer, M (2007) *Digital Copyright and the Consumer Revolution: Hands Off My IPod*. Cheltenham: Edward Elgar.

Risse-Kappen, T (1995) *Bringing Transnational Relations Back In: Non-State Actors, Domestic Structures and International Institutions*. Cambridge: Cambridge University Press.

Roderick, L (2014) 'Discipline and power in the digital age: the case of the us consumer data broker industry' *Critical Sociology* 40(5): 729–46.

Rogers, M and Eden G (2017) 'The Snowden disclosures, technical standards and the making of surveillance infrastructures' *International Journal of Communication* 11: 802–23.

Rosenberg, J, Schulzrinne, H, Camarillo, G, Johnston, A, Peterson, J, Sparks, R, Handley, M, and Schooler, E (2002) 'SIP: Session Initiation Protocol' RFC 3261, June <http://www.rfc-editor.org/info/rfc3261>.

Roskind, J (2012) 'Multiplexed stream transport over UDP' <https://docs.google.com/document/d/1RNHkx_VvKWyWg6Lr8SZ-saqsQx7rFV-ev2jRFUoVD34/edit>.

Russell, A (2011) 'Constructing legitimacy: the W3C's patent policy' in L DeNardis (ed.), *Opening Standards: The Global Politics of Interoperability*. Boston. MA: MIT Press, pages 159–176.

Ronie, G-J (2000) *Exporting Environmentalism: U.S. Multinational Chemical Corporations in Brazil and Mexico*. Cambridge, MA: MIT Press.

Rüth, J, Poese, I, Dietzel, C, and Hohlfeld, O (2018) 'Passive and active measurement. A first look at QUIC in the wild'. International Conference on Passive and Active Network Measurement PAM, pp 255–68. https://arxiv.org/abs/1801.05168.

Ryszard Struzak and Dariusz Wiecek (2013) 'Regulatory Issues for TV White Spaces' in E. Pietrosemoli and M. Zennaro (Eds) *TV White Spaces – A Pragmatic Approach*, National Institute of Telecommunications Poland (ICTP), December, pages 19–34, http://wireless.ictp.it/tvws/book/.

Safavi-Naini, R and Nichols PS (2010) 'Digital rights management' in B Rosenberg (ed.), *Handbook of Financial Cryptography and Security*. London: Chapman & Hall, pp. 193–220.

Salter, I (1999) 'The standards regimes for communication and information technologies' in AC Cutler, V Haufler, and T Porter (eds), *Private Authority and International Affairs*. Albany, NY: SUNY Press.

Sarker, Z (2018) 'QUIC—a vehicle for transport protocol evolution' *Ericsson Research Blog* 4 June. https://www.ericsson.com/en/blog/2018/6/quic--a-vehicle-for-transport-protocol-evolution.

Sharpe, N and Olufunmilayo, A (2007) 'Is Apple playing fair? Navigating the iPod FairPlay DRM controversy' *Northwestern Journal of Technology and Intellectual Property* 5(2): 332–50.

Schellingerhout, R (2011) 'Standard-setting from a competition law perspective' *Competition Policy Newsletter. European Commission 2011–1*, https://ec.europa.eu/competition/publications/cpn/2011_1_1_en.pdf.

Schepel, H (2005) *The Constitution of Private Governance*. London: Bloomsbury Press.

Schmitter, P (2000) *How to Democratize the European Union . . . And Why Bother?* Lanham: Rowman & Littlefield Publishers.

Scholte, JA (2000) 'Global civil society' in N Woods (ed.), *The Political Economy of Globalization*. London: Palgrave. pp. 192–3.

Scholte, JA (2002) 'Civil society and democracy in global governance' *Global Governance* 8(3): 281–304.

Scholte, JA (2007) 'Civil society and the legitimation of global governance' *Journal of Civil Society* 3(3): 305–26.

Scholte, JA (2010) 'Governing a more global world' *Corporate Governance* 10(4): 459–74.

Scholte, JA (2017) 'Polycentrism and democracy in Internet governance' in U Kohl (ed.), *The Net and the Nation State Multidisciplinary Perspectives on Internet Governance*. Cambridge: Cambridge University Press, pages 165–84.

Scholte, J (2000) *Globalization: A Critical Introduction*. London: St. Martin's Press.

Scholte, J (2002) 'Civil society and democracy in global governance' *Global Governance* 8(3): 281–304.

Scholte, J (2007) 'Civil society and the legitimation of global governance' *Journal of Civil Society* 3: 305–26.

Sikkink, K (2011) *The Justice Cascade: How Human Rights Prosecutions Are Changing World Politics*. New York, NY: W.W. Norton & Company.

Simcoe, T (2007) 'Delay and *de jure* standardization: exploring the slowdown in Internet standards development' in S Greenstein and V Stango (eds), *Standards and Public Policy*. Cambridge: Cambridge University Press.

Simcoe, T (2014) 'Governing the anticommons: institutional design for standard-setting organizations' *Innovation Policy and the Economy* 14(1): 99–128.

Sivonen, H (2007) 'An HTML5 Conformance Checker' Master's Thesis submitted in partial fulfillment of the requirements for the degree of Master of Science in Technology. University of Helsinki, 7 May 2007.

Sky (2013) 'TV White Spaces: approach to coexistence'. British Sky Broadcasting Group plc ('Sky') Response. December <https://www.ofcom.org.uk/consultations-and-statements/category-1/white-space-coexistence>.

Slaughter, A-M (2004) *A New World Order*. Princeton, NJ: Princeton University Press.

Slavka, A (2008) *Powerscape of Internet governance: how was global multistakeholderism invented in ICANN?* Riga: VDM Verlag.

Smith, K (2017) 'QUIC: a view from Mobile Operators' Presentation at Current Issues in Standards Development Organisations Decision-Making Workshop. 15 November 2017, Portcullis House, London.

Snellman, J (2016) 'The hidden cost of QUIC and TOU' 1 December 2016, weblog <https://www.snellman.net/blog/archive/2016-12-01-quic-tou/>.

Soghoian, C (2011) 'Slight paranoia: the history of the do not track header' 21 January 2011 <http://paranoia.dubfire.net/2011/01/history-of-do-not-track-header.html>.

Stanley, D, Montemurro, M, and Gilb, J (2016) 'SIG impact on the IEEE 802.11ax TG—IEEE Standards Association' Report: 802.11ax dominance complaint, 9 November, doc.: IEEE 802.11-16/1519r0 <https://mentor.ieee.org/802.11/dcn/16/11-16-1519>.

Statista (2019) Global Wi-Fi enabled equipment shipments 2012–2017 <https://www.statista.com/statistics/282512/global-wi-fi-enabled-equipment-shipments/>.

Steffek, J Kissling, C, and Nanz, P (2008) *Civil Society Participation in European and Global Governance A Cure for the Democratic Deficit?* London: Palgrave.

Steffek, J (2010) 'Public accountability and the public sphere of international governance' *Ethics & International Affairs* 24(1): 45–68.

Stephens A (2016) 'Dominance allegation in TGax' IEEE P802.11, 16 June, doc.: IEEE 802.11-16/0784r0, https://mentor.ieee.org/802.11/dcn/16/11-16-0784-00-0000-dominance-allegation-in-tgax.doc.

Stephens, A (2016) 'Entity submissions related to TGax special measures' IEEE p802.11 Wireless LANs, doc.: IEEE 802.11-16/1575r7, 16 December, https://mentor.ieee.org/802.11/dcn/16/11-16-0770-01-0000-july-2016-802-11-opening-report.pptx.

Stephens, A (2017) '802.11 Working Group Opening Report January 2017' doc.: IEEE 802.11-16/1561r0, 15 January.

Stone, D and Ladi, S (2015) 'Global public policy and transnational administration' *Public Administration* 93(4): 839–55.

Strange, S (1996) *The Retreat of the State: The Diffusion of Power in the World Economy*. Cambridge: Cambridge University Press.

Streeck, W and Thelen, K (eds) (2005) *Beyond Continuity: Institutional Change in Advanced Political Economies*, Oxford: Oxford University Press.

Tallberg, J, Sommerer, T, Squatrito T, and Lunds, CJ (2013) *The Opening Up of International Organizations Transnational Access in Global Governance*. Cambridge: Cambridge University Press.

Tarrow, S (1994) *Power in Movement: Social Movements, Collective Action and Politics*. New York, NY: Cambridge University Press.

techUK (2013) 'techUK response to the Ofcom consultation on TV white spaces: approach to coexistence' 23 December <https://www.ofcom.org.uk/consultations-and-statements/category-1/white-space-coexistence>.

Temperton, J (2015) 'TV white space will connect the internet of things' 13 February, <https://www.wired.co.uk/article/white-space-spectrum>.

Ten Oever, N (2018) 'Up to whose standards? Privacy, freedom & human rights' *IEEE Standards University* 9 (3), https://www.standardsuniversity.org/issue/october-2018-volume-9-issue-3-privacy-freedom-human-rights/.

Ten Oever, N and Beraldo D (2018) 'Routes to rights: Internet architecture and values in times of ossification and commercialization' *XRDS: Crossroads, The ACM Magazine for Students—Pseudonimity and Anonymity* 24(4): Summer, 28–31.

[TLS] Industry Concerns about TLS 1.3 (2016), IETF Mail archive: https://mailarchive.ietf.org/arch/msg/tls/CzjJB1g0uFypY8UDdr6P9SCQBqA.

Toner, A, Eckersley, P, and Reitman, R. (2015), 'Clear Rules of the Road with the Do Not Track Policy', Electronic Frontiers Foundation, https://www.eff.org/deeplinks/2015/08/clear-rules-road-do-not-track-policy-0.

Trondal, J Marcussen, M, Larsson, T, and Veggeland, F (2012) 'European administrative centre formation. Lessons from international bureaucracies' *Comparative European Politics* 10(1): 86–110.

Van Eeten, MJG and Mueller M (2013) 'Where is the governance in Internet governance' *New Media & Society* 15(5): 720–36.

Vibert, F (2007) *The Rise of the Unelected: Democracy and the New Separation of Powers*. Cambridge: Cambridge University Press.

Vodafone (2013) 'Vodafone's response to Ofcom Consultation TV white spaces: approach to coexistence' November <https://www.ofcom.org.uk/consultations-and-statements/category-1/white-space-coexistence>.

Vohra, N (2016) 'Email response from Broadcom to 802.11 Chair' document ec-16-0198-00-00EC, 16 November <https://mentor.ieee.org/802-ec/dcn/16/ec-16-0198-00-00EC-response-to-dominance-allegation-recommendation.docx>.

Voice of the Listener and Viewer (2013) 'Response from the Voice of the Listener and Viewer to Ofcom's Consultation on TV white spaces: approach to coexistence' 13 December <https://www.ofcom.org.uk/consultations-and-statements/category-1/white-space-coexistence>.

Warren, T (2017) 'Microsoft wants to close the rural broadband gap with TV white spaces', *The Verge*, 11 July, https://www.theverge.com/2017/7/11/15953310/microsoft-rural-airband-broadband-strategy.

Weightless Special Interest Group (2013) 'Weightless response to Ofcom Consultation "TV white spaces: approach to co-existence"' <https://www.ofcom.org.uk/consultations-and-statements/category-1/white-space-coexistence>..

Weston, S and Kretschmer, M (2012) 'Open standards in Government IT: a review of the evidence'. London: Cabinet Office.

Wi-Fi Alliance (2013) 'Wi-Fi Alliance response to Ofcom consultation: TV white spaces: approach to coexistence' 13 December <https://www.ofcom.org.uk/consultations-and-statements/category-1/white-space-coexistence>.

Wireless Broadband Alliance (2015) 'Wireless Broadband Alliance (WBA) response to Federal Communications Commission (FCC) Office of Engineering of and Technology and Wireless Telecommunications Bureau public seek of Information on Current Trends in LTE-U and LAA Technology' ET Docket No. 15-105. 29/06/15 <https://www.fcc.gov/ecfs/filing/60001086066/document/60001105679>.

World Economic Forum (2016) Where is the world's Wi-Fi? <https://www.weforum.org/agenda/2016/03/where-are-the-world-s-wifi-networks/>.

W3C (2011) 'W3C Workshop on Web Tracking and User Privacy' 28/29 April 2011, Princeton University, NJ <http://www.w3.org/2011/track-privacy/report.html>.

W3C (2013) Encrypted Media Extensions. W3C Working Draft 10 May 2013 <https://dvcs.w3.org/hg/html-media/raw-file/tip/encrypted-media/encrypted-media-fpwd.html>.

W3C (2016) Information about W3C and Encrypted Media Extensions (EME) March 2016 <https://www.w3.org/2016/03/EME-factsheet.html>.

Yates, J and Murphy, CN (2019) 'Development of the W3C WebCrypto API standard 2012 to 2017' in *Engineering Rules: Global Standard Setting since 1880*, edited by J Yates and CN MurphyBaltimore, MD: John Hopkins University Press, pp. 269–92.

Vifell, ÅC (2010) 'WTO and the environmental movement: on the path to participatory governance?' in CJJ Tallberg (ed.), *Transnational Actors in Global Governance*. London: Palgrave Macmillan, pp 110–33.

Zahariadis, N (1992) 'To sell or not to sell? Telecommunications policy in Britain and France' *Journal of Public Policy* 12(4): 355–76.

Zahariadis, N (2003) *Ambiguity and Choice in Public Policy: Political Decision Making in Modern Democracies*. Washington DC: Georgetown University Press.

Zahariadis, N (2005) *Essence of Political Manipulation: Emotion, Institutions, & Greek Foreign Policy*. New York: Peter Lang.

Zahariadis, N (2007) 'The multiple streams framework: structure, limitations, prospects' in PA Sabatier (ed.), *Theories of the Policy Process*. Boulder, CO: Westview Press, pp. 64–92.

Zahariadis, N (2013) 'Building better theoretical frameworks of the European Union's policy process' *Journal of European Public Policy* 20(6): 807–16.

Zahariadis, N (2016) 'Delphic oracles: ambiguity, institutions, and multiple streams' *Policy Sciences* 49(1): 3–12.

Zahariadis, N (ed.) (2016) *Handbook on Public Policy Agenda Setting*. Cheltenham: Edward Elgar.

Ziewitz, M and Pentzold C (2014) 'In search of internet governance: performing order in digitally networked environments' *New Media & Society* 16(2): 306–22.

Index